Charge and Energy Storage in Electrical Double Layers

Interface Science and Technology

Series Editor: Arthur T. Hubbard

In this Series:

Vol. 1: Clay Surfaces: Fundamentals and Applications
Edited by F. Wypych and K.G. Satyanarayana

Vol. 2: Electrokinetics in Microfluids
By Dongqing Li

Vol. 3: Radiotracer Studies of Interfaces
Edited by G. Horányi

Vol. 4: Emulsions: Structure Stability and Interactions
Edited by D.N. Petsev

Vol. 5: Inhaled Particles
By Chiu-sen Wang

Vol. 6: Heavy Metals in the Environment: Origin, Interaction and Remediation
Edited by H.B. Bradl

Vol. 7: Activated Carbon Surfaces in Environmental Remediation
Edited by Teresa J. Bandosz

Vol. 8: Tissue Engineering: Fundamentals and Applications
By Yoshito Ikada

Vol. 9: Particles at Interfaces: Interactions, Deposition, Structure
By Zbigniew Adamczyk

Vol. 10: Interface Science in Drinking Water Treatment: Theory and Applications
Edited by G. Newcombe and D. Dixon

Vol. 11: Surface Complexation Modelling
Edited by Johannes Lützenkirchen

Vol. 12: Theory of Colloid and Interfacial Electric Phenomena
By Hiroyuki Ohshima

Vol. 13: Sorbent Deformation
By A.V. Tvardovskiy

Vol. 14: Advanced Chemistry of Monolayers at Interfaces: Trends in Methodology and Technology
Edited by Toyoko Imae

Vol. 15: Macromolecules in Solution and Brownian Relativity
By Stefano A. Mezzasalma

Vol. 16: The Properties of Water and their Role in Colloidal and Biological Systems
By Carel Jan van Oss

Vol. 17: Chemistry on Modified Oxide and Phosphate Surfaces: Fundamentals and Applications
By Robson Fernandes de Farias

Vol. 18: Interface Science and Composites
By Soo-Jin Park and Min-Kang Seo

Vol. 19: Nanoparticle Technologies: From Lab to Market
By Farid Bensebaa

Vol. 20: Particles at Interfaces
By Zbigniew Adamczyk

Vol. 21: Self-Assembly Processes at Interfaces: Multiscale Phenomena
By Vincent Ball

Vol. 22: Rheology of Emulsions: Electrohydrodynamics Principles
By Aleksandar M. Spasic

Vol. 23: Tailored Thin Coatings for Corrosion Inhibition using a Molecular Approach
By Simo Olavi Pehkonen and Shaojun Yuan

Interface Science and Technology
Volume 24

Charge and Energy Storage in Electrical Double Layers

Edited by
Silvia Ahualli
Department of Applied Physics,
University of Granada,
Granada, Spain

Ángel V. Delgado
Department of Applied Physics,
University of Granada,
Granada, Spain

ELSEVIER

ACADEMIC PRESS
An imprint of Elsevier

Academic Press is an imprint of Elsevier
125 London Wall, London, EC2Y 5AS, United Kingdom
The Boulevard, Langford Lane, Kidlington, Oxford OX5 1GB, United Kingdom
50 Hampshire Street, 5th Floor, Cambridge, MA 02139, United States
525 B Street, Suite 1650, San Diego, CA 92101, United States

ISBN: 978-0-12-811370-7
ISSN: 1573-4285

For information on all Academic Press publications
visit our website at https://www.elsevier.com/books-and-journals

Publisher: Susan Dennis
Acquisition Editor: Anita A. Koch
Editorial Project Manager: Amy M. Clark
Production Project Manager: Prem Kumar Kaliamoorthi
Cover Designer: Victoria Pearson

Typeset by SPi Global, India

Contents

Contributors ix

Section A
Fundamentals of the Electrical Double Layer

1. Charge and Potential Distribution in the Electrical Double Layer of Porous Materials: Models 3

Ángel V. Delgado, Silvia Ahualli, María M. Fernández, Guillermo R. Iglesias, and María L. Jiménez

1.1 Introduction 3
1.2 The Classical Gouy-Chapman Theory 4
1.3 The Gouy-Chapman-Stern-Grahame Model 6
1.4 The Case of Cylindrical Pores 8
1.5 The Modified Donnan Model 12
1.6 Simulations 13
1.7 Conclusions 16
References 16

2. Kinetics of Ion Transport in a Porous Electrode 19

Silvia Ahualli, María L. Jiménez, and Ángel V. Delgado

2.1 Introduction 19
2.2 Basic Physics of the Problem 20
2.3 Equations and Boundary Conditions 21
2.4 Numerical Solution: Concentration and Potential Profiles With Time 24
2.5 Implications for Energy Production 26
2.6 Capacitive Deionization Cycles 28
2.7 Membrane Capacitive Deionization Cycles 31
2.8 Conclusions 36
Acknowledgments 36
References 36

3. The Electrical Double Layer as a Capacitor. Evaluation of Capacitance in Different Solutions: Effect of Ion Concentrations, Sizes, and Valencies 39

María L. Jiménez, Silvia Ahualli, and María M. Fernández

3.1 Introduction 39

3.2 Theory 41
3.2.1 The surface charge and potential 41
3.2.2 The mD approach 42
3.2.3 The excess chemical potential 43
3.3 EDL Structure: Bikerman Equation vs C-S Model 44
3.4 Ion Specificity 47
3.4.1 Effect of valency 47
3.4.2 Effect of the ionic size 48
3.5 EDL Overlap 48
3.6 Multiionic Solutions 51
3.6.1 Competition between ions with different valency 51
3.6.2 Competition between ions with equal valency 53
3.6.3 Size effects: Beyond steric repulsion 56
3.7 Steric Effect in CDLE 56
Acknowledgments 60
References 60

Section B
Materials

4. Selection of Carbon Electrode Materials 65

Xingtao Xu, Junfeng Li, Yuquan Li, Bing Ni, Xinjuan Liu, and Likun Pan

4.1 Introduction 65
4.2 Carbons 66
4.2.1 Carbon aerogels 66
4.2.2 CNTs 68
4.2.3 Mesoporous carbon 70
4.2.4 Graphene 73
4.3 Nitrogen-Doped Carbons 73
4.3.1 Nitrogen-doped CAs 73
4.3.2 Nitrogen-doped graphene 77
4.3.3 Nitrogen-doped porous carbon spheres 79
References 81

Section C
Capacitive Energy Extraction From Double Layer Expansion (CDLE)

5. Capacitive Energy Extraction From Double Layer Expansion (CDLE). Fundamentals of the Method 87

Doriano Brogioli and Fabio La Mantia

5.1 Salinity Gradient Power 87
5.2 What Happens to the EDL When We Decrease the Concentration of the Solution? 92

5.3 CapMix Cycle 94
5.4 Theory of Salt Adsorption 96
5.5 Thermodynamics of the CapMix Process 102
5.6 Voltage Rise of the Electrodes 104
5.6.1 Voltage rise versus base voltage graph 104
5.6.2 Simple activated carbon 106
5.6.3 Activated carbon with surface modifications 107
5.7 Kinetic Aspects of CDLE 110
5.8 Comparison With Other SGP Techniques 112
References 114
Further Reading 117

6. Capacitive Energy Extraction From CDLE: Implementation 119

Guillermo R. Iglesias and María M. Fernández

6.1 Introduction 119
6.2 The Capmix Cell 120
6.3 Suggestions for Methodology 122
6.4 Optimum Working Conditions 124
6.5 Temperature Influence on CDLE 127
6.6 Carbon Materials Requirements 130
6.7 Salinity Difference Effects 133
6.8 Stackings of Capmix Cells 135
6.9 Conclusions 138
References 139

Section D
Capacitive Energy Extraction by Donnan Potential (CDP)

7. CDP: Fundamentals of the Method 143

Silvia Ahualli, María M. Fernández, Ángel V. Delgado, and Guillermo R. Iglesias

7.1 Introduction 143
7.2 Details of the CDP Method 144
7.3 Theory: Steady State 146
7.4 "Soft" Electrodes 149
7.5 Experimental Implementation 152
7.6 Improvement: Wire Electrodes 156
7.7 Increasing the Power: Additional Charging 157
7.8 Increasing the Power: Stacking of Wire Electrodes 161
7.9 Efficiency 161
7.10 Conclusions 164
References 164

Section E
Capacitive Deionization (CDI)

8. Principles and Theoretical Models of CDI: Experimental Approaches 169

Silvia Ahualli, Guillermo R. Iglesias, and Ángel V. Delgado

8.1 Introduction 169
8.2 Basic Principles 170
8.3 CDI Performance Evaluation 175
8.4 Membrane Capacitive Deionization 176
8.4.1 Basic principles 176
8.4.2 Internal structure 177
8.4.3 Practical issues 180
8.4.4 Further applications of MCDI 181
8.5 Treated Electrodes 182
8.6 Flow Electrodes 185
8.7 Conclusions 188
Acknowledgments 189
References 189

Section F
Conclusions

9. Future and Perspectives of the Capacitive Techniques 195

Silvia Ahualli and Ángel V. Delgado

9.1 Introduction 195
9.2 Limitations of First-Principles Approaches 196
9.3 Materials for Electrodes and Their Surface Treatment 196
9.4 Energy Balance: Where Are We Now? 197
9.5 Economic Feasibility 198
9.6 Conclusion: Perspectives for the Future 198
Acknowledgments 199
References 199

Index 203

Contributors

Numbers in Parentheses indicate the pages on which the author's contributions begin.

Silvia Ahualli (3, 19, 39, 143, 169, 195), Department of Applied Physics, University of Granada, Granada, Spain

Doriano Brogioli (87), Energy Storage and Energy Conversion Systems, University of Bremen, Bremen, Germany

Ángel V. Delgado (3, 19, 143, 169, 195), Department of Applied Physics, University of Granada, Granada, Spain

María M. Fernández (3, 39, 119, 143), Department of Applied Physics, University of Granada, Granada, Spain; Department of Fluidic and Energy Science, Central America University "José Simeón Cañas", San Salvador, El Salvador

Guillermo R. Iglesias (3, 119, 143, 169), Department of Applied Physics, University of Granada, Granada, Spain

María L. Jiménez (3, 19, 39), Department of Applied Physics, University of Granada, Granada, Spain

Junfeng Li (65), Shanghai Key Laboratory of Magnetic Resonance, School of Physics and Materials Science, East China Normal University, Shanghai, People's Republic of China

Yuquan Li (65), Shanghai Key Laboratory of Magnetic Resonance, School of Physics and Materials Science, East China Normal University, Shanghai, People's Republic of China

Xinjuan Liu (65), Institute of Coordination Bond Metrology and Engineering (CBME), China Jiliang University, Hangzhou, People's Republic of China

Fabio La Mantia (87), Energy Storage and Energy Conversion Systems, University of Bremen, Bremen, Germany

Bing Ni (65), Shanghai Key Laboratory of Magnetic Resonance, School of Physics and Materials Science, East China Normal University, Shanghai, People's Republic of China

Likun Pan (65), Shanghai Key Laboratory of Magnetic Resonance, School of Physics and Materials Science, East China Normal University, Shanghai, People's Republic of China

Xingtao Xu (65), Shanghai Key Laboratory of Magnetic Resonance, School of Physics and Materials Science, East China Normal University, Shanghai, People's Republic of China

Section A

Fundamentals of the Electrical Double Layer

Chapter 1

Charge and Potential Distribution in the Electrical Double Layer of Porous Materials: Models

Ángel V. Delgado, Silvia Ahualli, María M. Fernández, Guillermo R. Iglesias and María L. Jiménez
Department of Applied Physics, University of Granada, Granada, Spain

1.1 Introduction

It is almost 140 years since Hermann L.F. von Helmholtz [1] published his theory of the electrical double layer (EDL hereafter), the name given to the potential and charge arrangement existing at the, ideally two-dimensional, interface between a charged surface and an ionic solution. The Helmholtz model is now considered the simplest one, but it was the first approach to the existence of such charge distribution. The charge on the solid surface, of typically electronic origin in the case of conducting electrodes, is compensated for by ions with opposite charge to that of the surface (*counterions*) approaching the charged surface and forming another surface at some distance from it. The whole structure is hence a sort of capacitor with a plate separation comparable to the size of an ion, as depicted in Fig. 1.1. If we call σ_S the surface charge density of any of the layers, the electrical potential Ψ_S on the solid surface (with reference to the potential on the virtual surface in the solution) is given by:

$$\Psi_S = \frac{\kappa^{-1}}{\varepsilon\varepsilon_0}\sigma_S \tag{1.1}$$

where κ^{-1} is the plate-to-plate distance (the reason for this denomination will be clear shortly), ε_0 is the electric permittivity of vacuum, and ε is the relative permittivity of the solution.

But the first reference to a capacitor based precisely on the existence of EDLs goes back to just 50 years ago, when the first patent on supercapacitors or electrical-double-layer capacitors was filed (US Patent US2800616A,

Interface Science and Technology, Vol. 24. https://doi.org/10.1016/B978-0-12-811370-7.00001-2

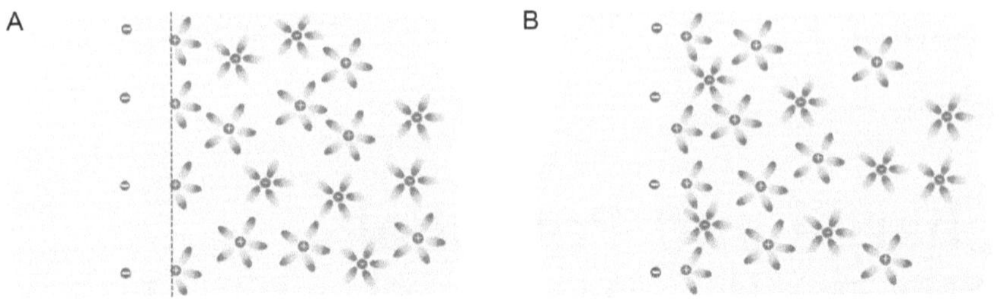

FIG. 1.1 The Helmholtz model of the EDL (A), and its present description (B). In the case of electrodes, the charge is of electronic origin. In other surfaces, charge is generated by ionized surface groups.

Low Voltage Electrolytic Capacitor; inventor: H.I. Becker, with General Electric Company). The reason for the name *supercapacitor* is of course because the capacitance of the EDL can be very high; supercapacitors with several thousand farads in a single unit weighing a few tens of grams are now commercially available. Their use as energy storage devices has become widespread in a number of technological applications. More recently, Brogioli [2] proposed for the first time that the changes in the capacitance of the EDL when the ionic contents of the solution are modified could be applied to the harvesting of clean electrical energy from salinity exchange, or, in modern terms, blue energy by CapMix (capacitive mixing) methods [3].

During the passing decades, the models of the charge-potential relationships in the EDL evolved with increased degrees of rigor and sophistication. In this chapter, we intend to provide the basics of the theories and methods available for evaluating the EDL properties, in fact the tools required for eventually predicting the energy that can be obtained from exchanging solutions in contact with the interfaces. The description is by no means as complete as that found in related monographs, such as the late J. Lyklema's volumes [4].

1.2 The Classical Gouy-Chapman Theory

The Helmholtz model neglected a fundamental contribution to the equilibrium configuration of the ions building the solution side of the EDL: if they accumulate close to the surface, a concentration gradient is established that acts against the accumulation. As a result, the mentioned part of the EDL is extended through a region of finite thickness, as first pointed out by Gouy [5] and Chapman [6]. In equilibrium, the electrostatic attraction of the charged wall on the counterions, and correspondingly the repulsion on the coions, together with the diffusion forces tending to equate concentrations, lead to a null gradient of electrochemical potential throughout the solution. If $z_j e$ is the charge of ionic species j ($j = 1,2,\ldots,N$), and c_j its concentration (in mol/m^3), then the condition of zero flux of any of the ions

$$\nabla\left[\mu_j^{\infty} + ez_j\Psi(\mathbf{r}) + k_{\mathrm{B}}T\,\ln c_j(\mathbf{r})\right] = 0 \tag{1.2}$$

where $\Psi(\mathbf{r})$ is the potential at an arbitrary position $\mathbf{r}$. Using the boundary conditions that, far from the interface, the concentration equals its bulk value c_j^0 and the potential tends to zero, we obtain the Boltzmann distribution for the concentration of each species, and the potential profile is obtained by solving Poisson's equation with a volume charge density given by the summation of the point charges in any volume element. The result is the Poisson-Boltzmann equation, the basis of the Gouy-Chapman model of the EDL:

$$\nabla^2\Psi(\mathbf{r}) = -\frac{eN_A}{\varepsilon\varepsilon_0}\sum_{j=1}^{N} z_j c_j^0 \exp\left(-ez_j\Psi(\mathbf{r})/k_{\mathrm{B}}T\right) \tag{1.3}$$

This partial differential equation typically needs numerical methods for its solution, but it can be solved analytically in a number of cases. Some of them are mentioned here [7]:

✓ *Flat interface and z-z electrolyte* ($N = 2$, $z_1 = -z_2 = z$ x axis perpendicular to the interface and with origin on it):

$$\Psi(x) = \frac{2k_{\mathrm{B}}T}{e}\ln\left[\frac{1+\gamma\exp(-\kappa x)}{1-\gamma\exp(-\kappa x)}\right]$$
$$\gamma = \tanh\left(\frac{ez\Psi_S}{4k_{\mathrm{B}}T}\right) \tag{1.4}$$

where κ is the reciprocal of the Debye length or EDL thickness, determined by the valences and concentrations of ions in solution:

$$\kappa^{-1} = \left(\frac{\varepsilon\varepsilon_0 k_{\mathrm{B}}T}{N_A\sum_{j=1}^{N} z_j^2 e^2 c_j^0}\right)^{1/2} =$$
$$\{\text{if}\, N=2,\ z_1 = -z_2 = z,\ c_1^0 = c_2^0 = c^0\}$$
$$= \left(\frac{\varepsilon\varepsilon_0 k_{\mathrm{B}}T}{2N_A z^2 e^2 c^0}\right)^{1/2} \tag{1.5}$$

Analytical solutions do also exist if the charged surface is in contact with a 2-1 electrolyte solution (see, e.g., [8]). Note here that the surface charge density can be easily expressed analytically. Since the charge is only located on the surface of the particles, the electric field inside them is zero, and the condition of continuity of the normal component of the displacement vector reads:

$$\varepsilon\varepsilon_0 E(x=0^+) = -\varepsilon\varepsilon_0 \left.\frac{d\Psi}{dx}\right|_{x=0^+} = \sigma_S \tag{1.6}$$

and using Eqs. (1.4) and (1.6), the charge density can be obtained:

$$\sigma_S = \frac{2\kappa k_B T \varepsilon\varepsilon_0}{ze} \sinh \frac{ze\Psi_S}{2k_B T} \tag{1.7}$$

✓ *Flat interface and low surface potential $e\Psi_S/k_BT \ll 1$ (Debye-Hückel approximation)*:

$$\Psi(x) = \Psi_S \exp(-\kappa x) \tag{1.8}$$

$$\sigma_S = \frac{\varepsilon\varepsilon_0}{\kappa^{-1}} \Psi_S \tag{1.9}$$

that is, the specific capacitance ($\varepsilon\varepsilon_0/\kappa^{-1}$) of the EDL would in this case correspond to that of a parallel plate capacitor with plates separated by a Debye length and containing a dielectric of relative permittivity ε.

✓ *EDL around a spherical particle of radius a*

The Poisson-Boltzmann equation (1.3) in spherical coordinates (r, θ, φ) becomes, considering the spherical symmetry:

$$\frac{d^2\Psi(r)}{dr^2} + \frac{2}{r}\frac{d\Psi(r)}{dr} = -\frac{eN_A}{\varepsilon_m} \sum_{j=1}^{N} z_j c_j^0 \exp\left(-ez_j \Psi(r)/k_B T\right) \tag{1.10}$$

As before, approximate solutions have been found in simple cases. For low surface potential,

$$\Psi(r) = \Psi_S \frac{a}{r} \exp[-\kappa(r-a)] \tag{1.11}$$

and the general solution must be numerical, although approximate expressions with negligible error in most circumstances have been found [8]. Fig. 1.2 illustrates the differences between full numerical calculations and the low-potential approximation. The agreement at low surface potentials and for the larger particle diameter is clear.

1.3 The Gouy-Chapman-Stern-Grahame Model

Lyklema [4] discussed in great detail the main failuers of the Gouy-Chapman model, including neglecting the finite volume of the ions and their mutual interactions, considering only electrostatic (and not specific or image forces) interactions between the ions and the charged surface (assumed to be flat and homogeneously charged), or identifying the electrical permittivity inside

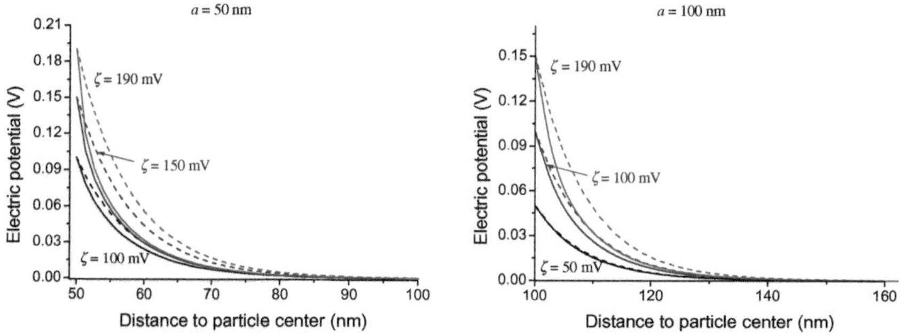

FIG. 1.2 The electrical potential profile in the diffuse layer around a spherical colloidal particle for different surface potentials. Dashed lines: approximation Eq. (1.11); solid lines: numerical calculation. Solution: 1 mM KCl; radius: 50 nm (left) and 100 nm (right).

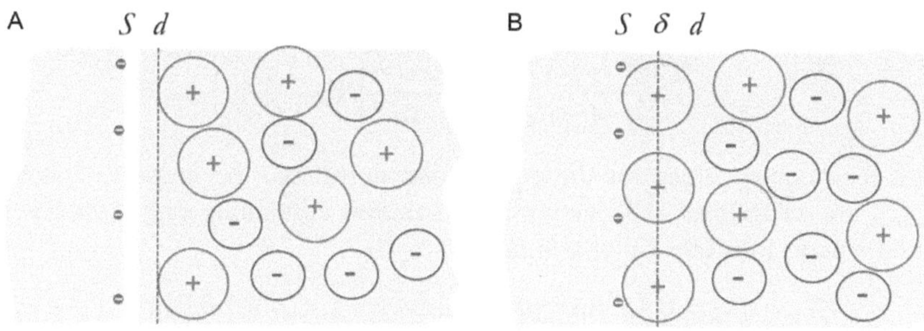

FIG. 1.3 The Gouy-Chapman-Stern-Grahame model. (A) The Stern layer is free of charge; (B) the layer is charged by specific adsorption of ions (counterions in the case illustrated).

the EDL and in the bulk, to mention a few. The first attempt to improve on these limitations was achieved by Otto Stern [9] who considered that the finite size of ions prevents them from approaching to the surface beyond their radius, leaving a charge-free region (Stern layer) with a thickness comparable to the ion radius (Fig. 1.3A). The potential falls linearly from the surface to the limit of the layer, known as the *outer Helmholtz plane* (OHP), which marks the beginning of the diffuse atmosphere at a distance d from the surface. The potential on the OHP is called diffuse or Stern potential, Ψ_d. The electroneutrality of the EDL now requires that the surface charge density be compensated for by an opposite equivalent surface charge on the OHP:

$$\sigma_S + \sigma_d = 0 \tag{1.12}$$

and an integral capacitance per unit area given by

$$C_{St} = \frac{|\sigma_S|}{|\Psi_S - \Psi_d|} \tag{1.13}$$

can be defined. This capacitance is often given per unit volume in the case of porous materials, as will be discussed shortly. In addition, there is the possibility that ions are specifically adsorbed at some distance δ from the solid surface, providing an additional charged layer, ideally located in the *inner Helmholtz plane* (IHP), where the potential is denoted Ψ_δ, and the distance to the surface is δ (Fig. 1.3B). The charge here is considered immobile under the action of external fields, but contributes to the charge balance, so that Eq. (1.12) must be modified to include this new charge density:

$$\sigma_S + \sigma_\delta + \sigma_d = 0 \tag{1.14}$$

The value of the inner-layer charge density can be estimated assuming a Langmuir-type adsorption isotherm [10, 11], meaning that a certain density of adsorption sites (N_j) are available for each ionic species ($j = 1,\ldots, N$), with dissociation constants K_j. The surface charge density σ_δ can thus be calculated as:

$$\sigma_\delta = \sum_{j=1}^{N} \frac{z_j e N_j \frac{c_j^0}{K_j} \exp\left(-\frac{z_j e}{k_B T}\left[\Psi_d - \frac{\sigma_d}{C_{St}}\right]\right)}{1 + \sum_{k=1}^{N} \frac{c_k^0}{K_k} \exp\left(-\frac{z_k e}{k_B T}\left[\Psi_d - \frac{\sigma_d}{C_{St}}\right]\right)} \tag{1.15}$$

where reasonable values for the equilibrium constants, the density of sites, and C_{St} are as follows [10], assuming a solution containing only two types of monovalent ions ($N = 2$; $z_1 = -z_2 = 1$):

$$eN_1 = eN_2 = 0.1\,\mathrm{C/m^2};\ \mathrm{p}K_1 = 2;\ \mathrm{p}K_2 = 1;\ C_{St} = 1.3\,\mathrm{F/m^2}$$

Using these values, and a Stern potential and diffuse charge density equal to $\Psi_d = -100$ mV, and $\sigma_d = 0.01$ C/m^2, respectively, we obtain, for a 1 mM solution, $\sigma_\delta = 0.09$ C/m^2, meaning that a significant portion of the double-layer charge is associated with the Stern layer. This has a number of implications in explaining the electric and electrokinetic behavior of interfaces with nonzero Stern layer charge [12, 13].

1.4 The Case of Cylindrical Pores

This is obviously the situation that we are most interested in. Although a cylindrical pore is certainly just a model, it is a realistic one: no matter what the actual pore size distribution is, a charged cylinder is a good approximation to their true geometry. The Poisson-Boltzmann equation must first be modified to account for the cylindrical geometry. We call r the radial cylindrical coordinate, and take into account the symmetry of the pore. If it is long enough to neglect edge effects, the potential Ψ will depend only on r. With this, Eq. (1.3) reads:

$$\frac{d^2\Psi(r)}{dr^2} + \frac{1}{r}\frac{d\Psi(r)}{dr} = -\frac{eN_A}{\varepsilon\varepsilon_0}\sum_{j=1}^{N} z_j c_j^0 \exp\left(-ez_j\Psi(r)/k_B T\right) \tag{1.16}$$

Ohshima [14] has reported a simplified analytical solution of the previous equation subject to the usual boundary conditions (r is measured from the axis of the cylinder):

$$\left.\frac{d\Psi}{dr}\right|_{r=0}=0$$
$$\Psi(r=R)=\Psi_S \tag{1.17}$$

His solution assumes low-to-moderate surface potential, and this means that it will hardly be useful for our purposes, as the electrodes are subjected to potential differences of around 300–500 mV, which is much higher than the few tens of mV typically admitted for low-potential approximations. Nevertheless, the analytical expressions are useful to give us a feeling of the typical potential profiles in the pore. Up to a second order in the surface potential, the solution is quite accessible, and we reproduce it here for the interested reader. First of all, one assumes that only two types of ions, with valences $+z$ and $-z$, are in solution, and use a dimensionless version of the potential, $y(r)=ze\Psi(r)/k_BT$. Then [14]:

$$\frac{d^2y(r)}{dr^2}+\frac{1}{r}\frac{dy(r)}{dr}=\kappa^2\sinh y$$
$$y(R)=0 \tag{1.18}$$
$$y(0)=y_0=\frac{ze\Psi_0}{k_BT}$$

The solution can be easily implemented. To order y_0^2:

$$y^{(2)}(r)=y_0-\left(y_0-y_{ref}^{(2)}+\tanh y_{ref}^{(2)}\right)\left[1-\frac{I_0\left(\kappa r\sqrt{\cosh y_{ref}^{(2)}}\right)}{I_0\left(\kappa R\sqrt{\cosh y_{ref}^{(2)}}\right)}\right] \tag{1.19}$$

where

$$y_{ref}^{(2)}=y^{(1)}(0) \tag{1.20}$$

and

$$y^{(1)}(r)=y_0-\left(y_0-y_{ref}^{(1)}+\tanh y_{ref}^{(1)}\right)\left[1-\frac{I_0\left(\kappa r\sqrt{\cosh y_{ref}^{(1)}}\right)}{I_0\left(\kappa R\sqrt{\cosh y_{ref}^{(1)}}\right)}\right] \tag{1.21}$$

being

$$
\begin{aligned}
y_{ref}^{(1)} &= y^{(0)}(0) \\
y^{(0)}(r) &= y_0 \tanh y_0 \left[1 - \frac{I_0\left(\kappa r \sqrt{\cosh y_0}\right)}{I_0\left(\kappa R \sqrt{\cosh y_0}\right)}\right]
\end{aligned}
\tag{1.22}
$$

In all these equations, $I_0(x)$ is the zeroth-order modified Bessel function of the first kind [15].

A general solution can only be reached numerically. In our case, we used Matlab for solving Eq. (1.16), and this can be done in the general case, that is, without limitations concerning either the potential value or the combination of ions in solution. Fig. 1.4 illustrates the comparison. The analytical formula fails in providing results at the higher potentials tested (200 and 300 mV) for both pore diameters and concentrations tested. Nevertheless, it is very close to the numerical calculation in all cases where comparison is possible. As observed, for the lowest concentration (thicker double layers), the potential falls to a constant value at a distance from the wall around 50% of the radius. The decrease is faster for higher concentrations and wall potentials.

As mentioned, the high wall potentials of the pore can provoke an unphysically large accumulation of counterions in the vicinity of the surface.

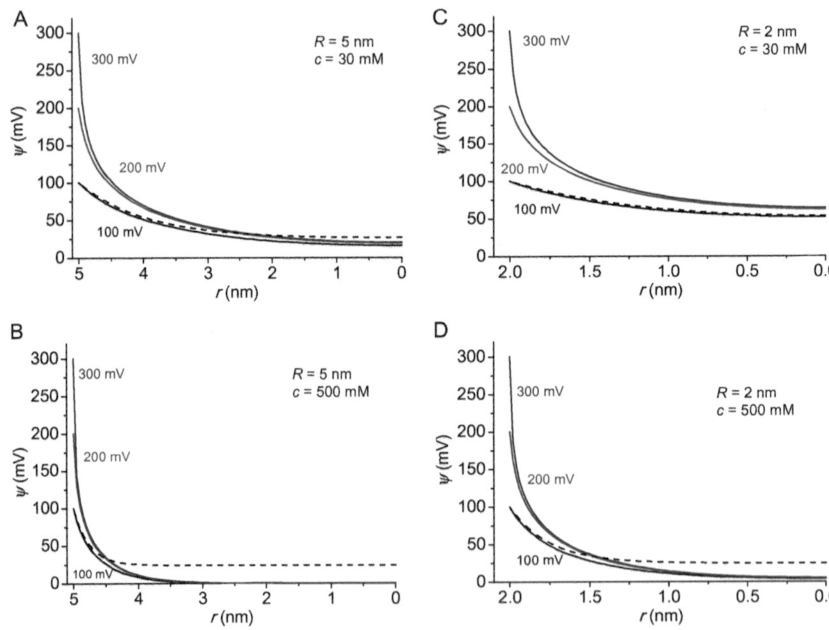

FIG. 1.4 Electric potential profiles in the cylindrical pore for different values of the pore radius and the wall potential. Solid lines: numerical solution; dashed lines: analytical approximation. (A, B): 30 mM NaCl; (C, D): 500 mM NaCl. ($r = 0$ corresponds to the centre of the pore).

The simplest way to minimize this effect is to prevent the ions from approaching in an unlimited fashion (only controlled by electrostatics and diffusion) to the surface of the pore. This can be done by taking into account that the ions have a finite size and cannot stay closer than their radius, *a* say (probably bare, as the hydration layer is likely to be lost, at least in the direction to the surface). This is equivalent to considering an uncharged Stern layer (Fig. 1.5), whose thickness is equal to *a*. Eq. (1.16) must be separately solved in the two regions [16]: for *r* between 0 and $(R-a)$ the equation will be fully solved subject to the boundary conditions:

$$\left.\frac{d\Psi}{dr}\right|_{r=0} = 0$$
$$\Psi(r = R - a^{+}) = \Psi_d \tag{1.23}$$

whereas in the charge free region close to the wall, we set $c_j = 0$, and boundary conditions:

$$\Psi(r = R - a^{-}) = \Psi(r = R - a^{+})$$
$$\left.\frac{d\Psi}{dr}\right|_{r=R-a^{-}} = \left.\frac{d\Psi}{dr}\right|_{r=R-a^{+}} \tag{1.24}$$

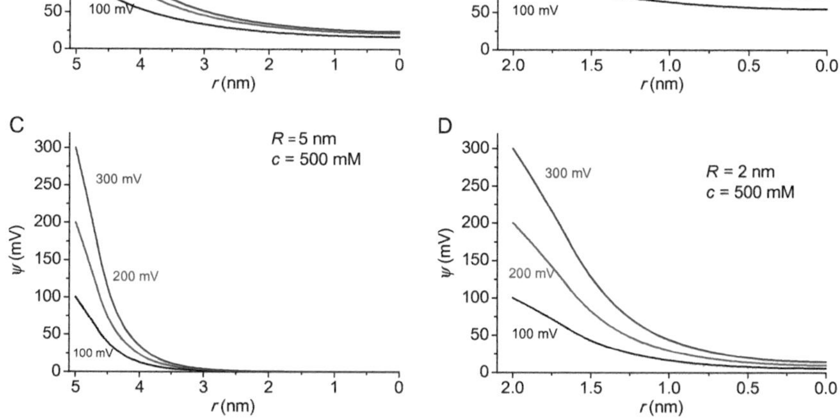

FIG. 1.5 The effect of the presence of a charge-free Stern layer on the potential dependence on radius for the cases considered in Fig. 1.4. Only numerical calculations are included.

The potential profile in the $(R-a,0)$ region will be linear (Fig. 1.5), so that alternatively, if the permittivity of the Stern layer is supposed to be equal to that of the bulk, we can define a cylindrical capacitor of length L and plate separation a whose capacitance will be:

$$C_{St} = \frac{2\pi\varepsilon\varepsilon_0 L}{\ln\left(\frac{R}{R-a}\right)} \simeq \varepsilon\varepsilon_0 \frac{S}{a} \tag{1.25}$$

where S is the plate area. In electrochemistry of porous media, it is usual to consider a Stern capacitance per unit volume of the cylindrical pore and not per unit area. The latter would be, according to Eq. (1.25), equal to $\varepsilon\varepsilon_0/a$, and the former $(\varepsilon\varepsilon_0 S/a)/(SL) = \varepsilon\varepsilon_0/(aL)$. If the average pore length is comparable to the electrode thickness, typically 100 μm, then the volume Stern-layer capacitance would be of the order of 10^4 F/m^3.

1.5 The Modified Donnan Model

As shown in Fig. 1.4, the potential profile in the pore is close to flat for the lowest concentrations of interest and the typical micropore diameters. This is taken advantage of in the so-called modified-Donnan model, widely used in approximate calculations of the capacitance of supercapacitors or electrodes for salinity exchange energy or deionization [17–19]. As long as the pore radius is smaller or comparable to the Debye length, the model can be reasonably applied. Instead of a potential profile, a single value of the potential at the EDL, known as the Donnan potential Ψ_{Don}, suffices for evaluating the potential drop at the diffuse EDL and the ion concentration in the micropore, $c_{j,mi}$:

$$c_{j,mi} = c_j^0 \exp\left(-z_j e \Psi_{Don}/k_B T\right) \tag{1.26}$$

and the reference concentration c_j^0 is interpreted as that in the macropores. Thus, from the pore wall, the total potential drop equals the sum of Stern and Donnan potentials [20]. Note the origin of the denomination: in the strong overlap conditions mentioned, the constancy of the potential reminds us of the potential inside polyelectrolyte layers or ion-specific membranes. Let us imagine an anion exchange membrane, that is, containing positive dissociated groups inside its structure. If the concentration of the groups is high enough, coions (cations) will be rejected and only anions will be allowed inside the membrane. This accumulation will stop when the electric field created by the anions prevents further ions from approaching: a potential difference will hence be established between the membrane interior and the external solution. If the distribution of charged groups is homogeneous, the potential inside the membrane will be constant. This is the Donnan potential [21, 22], which is lower the lesser the difference between the concentration of charged groups in the membrane and that of the external solution. The model is named *modified* because [20] it includes the presence of the Stern layer, as previously mentioned, and also some nonelectrostatic potential of interaction between ions from the macropores and the micropore walls.

1.6 Simulations

Because of the numerical complications associated with the GCS model in general conditions, computer simulations have been developed for progressing over numerical or analytical solutions, while offering new insights, not always appreciated in the latter. Monte Carlo (MC) simulations were first performed in the 1970s and early 1980s by Valleau and Torrie and coworkers [23–29]. These authors carried out MC calculations in the grand canonical ensemble in a square prism with the surface charge on one of its bases and no charge on the other, and periodic boundary conditions. The bulk electrolyte is described according to the primitive model: ions are spheres of specified size and charge and the solvent is a continuum characterized by its dielectric permittivity. In the case of 1-1 electrolyte [23] with concentration up to 1 M, the results of the MC simulations are in agreement with the GCS model (Fig. 1.6). The success of the latter model is by no means that significant when the ions have charges other than +1 or −1. In a later work [25], these authors found that for 2-2 and 2-1 electrolytes the GCS results overestimate the surface potential and predict oscillations in the counterion concentration and electric potential profiles that some authors consider responsible for charge reversal in electrophoresis [30–34].

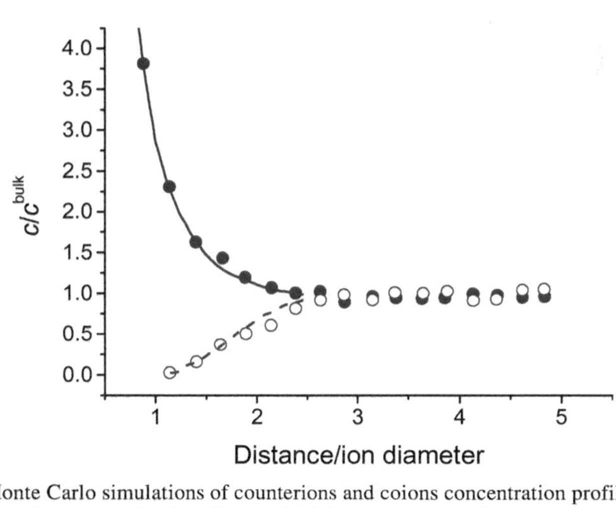

FIG. 1.6 Monte Carlo simulations of counterions and coions concentration profiles, compared to GCS calculations, for a flat interface in 1 M 1-1 electrolyte. The surface charge density is $-12.5\ \mu C/cm^2$. The lines are GCS predictions; the symbols are MC results for counterions (full symbols) and cooins (open symbols). *Reproduced with permission from Torrie GM, Valleau JP. Electrical double layers. 4. Limitations of the Gouy-Chapman theory. J Phys Chem 1982;86:343–346. © 1982, American Chemical Society.*

As shown in Fig. 1.7 [25], the potential changes sign from negative to positive at a distance comparable to the hard-sphere radius of the ions and displays a clear minimum at a distance equal to the ion diameter.

As mentioned, such phenomena of charge inversion in the EDL have been detected in an indirect fashion when measuring the electrophoretic mobility of particles as a function of the concentration of 2-1, 3-1, and 2-2 electrolyte solutions. Details of the approach are given, for instance, in [31–34]. Like in the MC approach, a short-ranged electrostatic interaction is assumed between hard-sphere ions, and no structure is used for the solvent. The Ornstein-Zernike formulation is used, with ion-ion correlations described by the *mean spherical approximation* (MSA), and the particle-ion interactions through the *hypernetted chain* (HNC) model. The resulting HNC/MSA integral equation is then numerically solved. A comparison carried out by Martín-Molina et al. [33] suggested that the MC simulations provided a better description of the electrophoretic mobility vs. concentration trends than the integral equation approach, for moderate to high salt concentrations.

Molecular dynamics [35] has also been used for evaluation of the ionic profiles in EDLs. The method is based on the application of classical mechanics laws to the time evolution of the discrete species in the test box, with proper interaction potentials and ranges, and boundary conditions at the box walls. Interestingly, it has been applied to the evaluation of blue energy, the main topic of the present book, and to the related field of supercapacitor design and evaluation [36, 37].

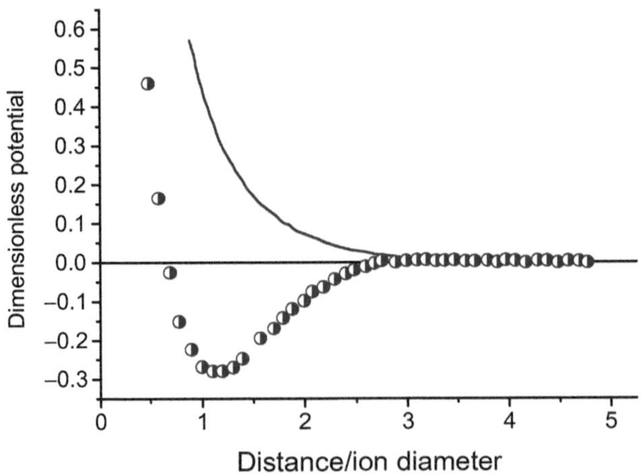

FIG. 1.7 Monte Carlo simulation of electric potential distribution, compared to GCS calculations, for a flat interface in 0.5 M 2-1 electrolyte. The surface charge density is $-15.1\ \mu C/cm^2$. The lines are GCS predictions; the symbols are MC results. *Reproduced with permission from Torrie GM, Valleau JP. Electrical double layers. 4. Limitations of the Gouy-Chapman theory. J Phys Chem 1982;86:343–346. © 1982, American Chemical Society.*

As an example, Fig. 1.8 illustrates the simulation system built in [37] for calculating the capacitance of an electrochemical cell consisting of two carbon electrodes of known porosity in contact with a sodium chloride solution (1 M) while charging for performing a Capmix energy harvesting cycle. Fig. 1.8 shows the system: 3821 carbon atoms constitute each electrode, together with 7615 water molecules and 70 Na and 70 Cl ions. Lennard-Jones and Coulomb

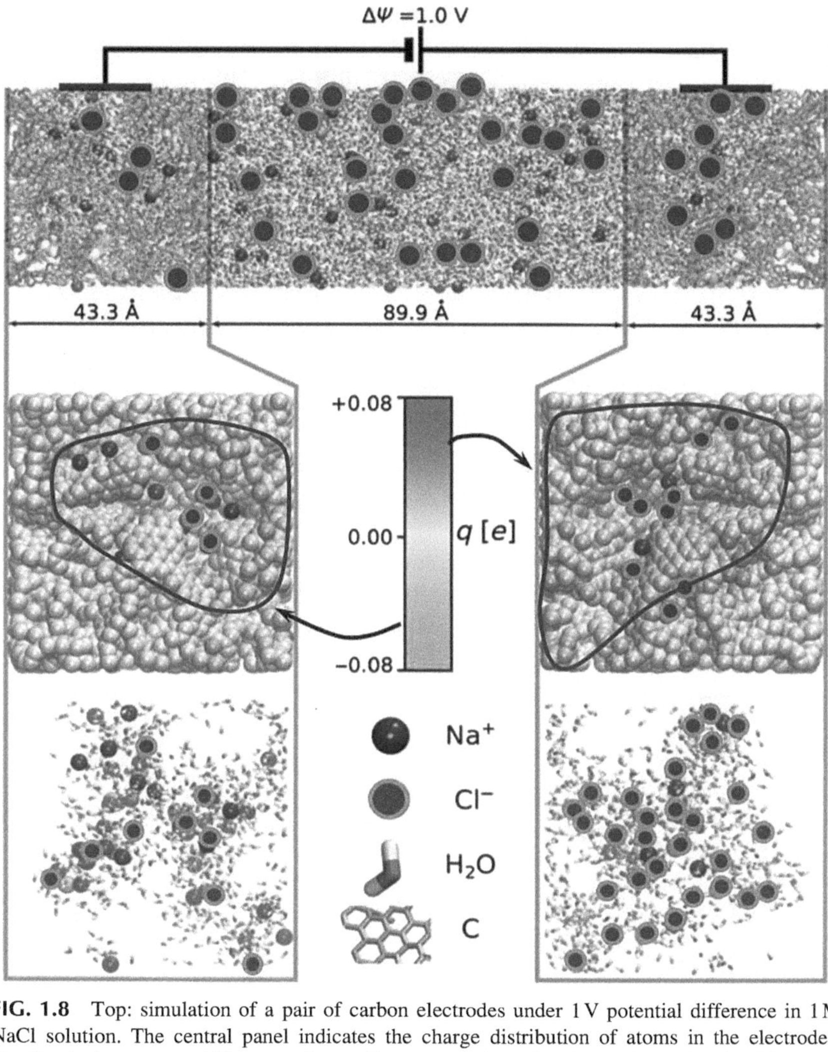

FIG. 1.8 Top: simulation of a pair of carbon electrodes under 1 V potential difference in 1 M NaCl solution. The central panel indicates the charge distribution of atoms in the electrodes. Finally, the bottom panel illustrates the confinement of electrolyte in the electrodes. *Modified with permission from Simoncelli M, et al.: Blue energy and desalination with nanoporous carbon electrodes: capacitance from molecular simulations to continuous models, Phys Rev X 8:021024, 2018. https://doi.org/10.1103/PhysRevX.8.021024. © 2018, American Physical Society.*

potentials are used to describe the interactions between ions. The Ewald summation criterion is used for the computation of electrostatic interactions. The success of the simulation is confirmed through comparison with experimental determinations in an electrochemical cell; for example, the capacitance at 1 M NaCl results 117 F/g in the simulation and 121 F/g in the experiments.

Most importantly, these simulations not only demonstrate their advantage over EDL calculations regarding their predictive power and design suggestions, but they also can provide precise explanations for data on energy storage and transport that are being collected continuously due to the appearance of new electrode materials currently available. As an example, Pak and Hwang [38] have demonstrated that the extreme confinement of ions in subnanometer pores, which are determinant of the high capacitance of electrical double layer capacitors, implies that the transport of ions is facilitated if the shape (and not just the size) of the pore is controlled. Thus, the capacitance is strongly correlated with the kinetics of ion reorganization inside the pores, which is in turn related to a new property, denominated *pore shape factor*. Dispersity in size and also in shape facilitates extraction of confined ions towards macropores where bulk-properties are prevalent.

1.7 Conclusions

The equilibrium structure of the electrical double layer is a classical topic of electrochemistry dating back to the early 20th century. Nevertheless, it is rather recently that it found application in the fields of energy harvesting and storage and deionization of aqueous solutions. In addition, progress in our understanding of the fine structure of the EDL continues to develop, particularly considering the ion-ion and ion-wall interactions mediated by the ion dimensions and charge. Although analytical and numerical methods have been envisaged, simulations are bringing new insight into the problem of the ion distribution inside the nanometer-size pores of the electrode materials used. Here is probably the answer for the problem of selecting such materials and improving the methods so as to maximize the performance of the energy or desalination devices. This will be dealt with in detail in the following chapters of this book.

References

[1] Helmholtz Hv. Studien über electrische Grenzschichten. Ann Phys 1879;243(7):337–82.

[2] Brogioli D. Extracting renewable energy from a salinity difference using a capacitor. Phys Rev Lett 2009;103(5):058501.

[3] Bijmans MFM, Burheim OS, Bryjak M, Delgado A, Hack P, Mantegazza F, et al. CAPMIX—deploying capacitors for salt gradient power extraction, technoport 2012. In: Sharing possibilities and 2nd renewable energy research conference (Rerc2012). 20:2012. p. 108–15.

[4] Lyklema J. Fundamentals of interface and colloid science. Vol. II. Londres: Academic Press; 1995.

[5] Gouy M. Sur la constitution de la charge électrique à la surface d'un électrolyte. J Phys Theor Appl 1910;9(1):457–68.
[6] Chapman DL. A contribution to the theory of electrocapillarity. Philos Mag 1913;25 (148):475–81.
[7] Ohshima H. Electrical phenomena at interfaces and biointerfaces. Hoboken, USA: John Wiley & Sons; 2012.
[8] Ohshima H, Furusawa K. Electrical phenomena at interfaces. New York: Marcel Dekker; 1998.
[9] Stern O. The theory of the electrolytic double shift. Zeitsch Elektrochem Angewand Physik Chem 1924;30:508–16.
[10] Mangelsdorf CS, White LR. Effects of stern-layer conductance on electrokinetic transport-properties of colloidal particles. J Chem Soc Faraday Trans 1990;86(16):2859–70.
[11] Mangelsdorf CS, White LR. The dynamic double layer—part 1—theory of a mobile Stern layer. J Chem Soc Faraday Trans 1998;94(16):2441–52.
[12] Carrique F, Arroyo FJ, Delgado AV. Electrokinetics of concentrated suspensions of spherical colloidal particles: effect of a dynamic Stern layer on electrophoresis and DC conductivity. J Colloid Interface Sci 2001;243(2):351–61.
[13] Jimenez ML, Arroyo FJ, Carrique F, Kaatze U, Delgado AV. Determination of stagnant layer conductivity in polystyrene suspensions: temperature effects. J Colloid Interface Sci 2005;281(2):503–9.
[14] Ohshima H. A simple algorithm for the calculation of the electric double layer potential distribution in a charged cylindrical narrow pore. Colloid Polym Sci 2016;294(11):1871–5.
[15] Abramowitz M, Stegun IA. Handbook of mathematical functions. Washington, DC: National Bureau of Standards; 1972.
[16] Ahualli S, Fernandez MM, Iglesias G, Delgado AV, Jimenez ML. Temperature effects on energy production by salinity exchange. Environ Sci Technol 2014;48(20):12378–85.
[17] Wang L, Biesheuvel PM, Lin SH. Reversible thermodynamic cycle analysis for capacitive deionization with modified Donnan model. J Colloid Interface Sci 2018;512:522–8.
[18] Zhao R, van Soestbergen M, Rijnaarts HHM, van der Wal A, Bazant MZ, Biesheuvel PM. Time-dependent ion selectivity in capacitive charging of porous electrodes. J Colloid Interface Sci 2012;384:38–44.
[19] Rica RA, Ziano R, Salerno D, Mantegazza F, Bazant MZ, Brogioli D. Electro-diffusion of ions in porous electrodes for capacitive extraction of renewable energy from salinity differences. Electrochim Acta 2013;92:304–14.
[20] Biesheuvel PM, Fu Y, Bazant MZ. Diffuse charge and Faradaic reactions in porous electrodes. Phys Rev E 2011;83(6):061507.
[21] Ohshima H, Kondo T. Membrane-potential and donnan potential. Biophys Chem 1988;29(3):277–81.
[22] Makino K, Ohshima H, Kondo T. Surface-potential of an ion-penetrable charged membrane. J Theor Biol 1987;125(3):367–8.
[23] Torrie GM, Valleau JP. Monte-Carlo study of an electrical double-layer. Chem Phys Lett 1979;65(2):343–6.
[24] Torrie GM, Valleau JP. Electrical double-layers. 1. Monte-carlo study of a uniformly charged surface. J Chem Phys 1980;73(11):5807–16.
[25] Torrie GM, Valleau JP. Electrical double-layers. 4. Limitations of the Gouy-Chapman theory. J Phys Chem 1982;86(16):3251–7.
[26] Torrie GM, Valleau JP, Patey GN. Electrical double-layers. 2. Monte-Carlo and HNC studies of image effects. J Chem Phys 1982;76(9):4615–22.

[27] Valleau JP, Torrie GM. The electrical double-layer. 3. Modified gouy-chapman theory with unequal ion sizes. J Chem Phys 1982;76(9):4623–30.
[28] Torrie GM, Valleau JP, Outhwaite CW. Electrical double-layers. 6. Image effects for divalent ions. J Chem Phys 1984;81(12):6296–300.
[29] Valleau JP, Torrie GM. Electrical double-layers. 5. Asymmetric ion wall interactions. J Chem Phys 1984;81(12):6291–5.
[30] Jimenez ML, Delgado AV, Lyklema J. Hydrolysis versus ion correlation models in electrokinetic charge inversion: establishing application ranges. Langmuir 2012;28(17):6786–93.
[31] Quesada-Perez M, Gonzalez-Tovar E, Martin-Molina A, Lozada-Cassou M, Hidalgo-Alvarez R. Overcharging in colloids: beyond the Poisson-Boltzmann approach. ChemPhysChem 2003;4(3): 234–48.
[32] Quesada-Perez M, Gonzalez-Tovar E, Martin-Molina A, Lozada-Cassou M, Hidalgo-Alvarez R. Ion size correlations and charge reversal in real colloids. Colloids Surf A Physicochem Eng Aspect 2005;267(1-3):24–30.
[33] Martin-Molina A, Quesada-Perez M, Hidalgo-Alvarez R. Electric double layers with electrolyte mixtures: integral equations theories and simulations. J Phys Chem B 2006;110(3): 1326–31.
[34] Martin-Molina A, Alberto Maroto-Centeno J, Hidalgo-Alvarez R, Quesada-Perez M. Charge reversal in real colloids: experiments, theory and simulations. Colloids Surf A Physicochem Eng Aspect 2008;319(1-3):103–8.
[35] Rapaport DC. The art of molecular dynamics simulations. Cambridge: Cambridge University Press; 2004.
[36] Bo Z, Li CW, Yang HC, Ostrikov K, Yan JH, Cen KF. Design of supercapacitor electrodes using molecular dynamics simulations. Nano-Micro Lett 2018;10(2):33.
[37] Simoncelli M, Ganfoud N, Sene A, Haefele M, Daffos B, Taberna P-L, et al. Blue energy and desalination with nanoporous carbon electrodes: capacitance from molecular simulations to continuous models. Phys Rev X 2018;8(2):021024.
[38] Pak AJ, Hwang GS. Molecular insights into the complex relationship between capacitance and pore morphology in nanoporous carbon-based supercapacitors. ACS Appl Mater Interfaces 2016;8(50):34659–67.

Chapter 2

Kinetics of Ion Transport in a Porous Electrode

Silvia Ahualli, María L. Jiménez and Ángel V. Delgado
Department of Applied Physics, University of Granada, Granada, Spain

2.1 Introduction

Although in Chapter 1 of this work, we have focused on steady-state charging properties of the EDLs in pores, applications in energy harvesting and storage and in deionization of aqueous electrolyte solutions also require consideration of the kinetics of the process. This is so, not only because power (and not just energy) produced or invested is a fundamental quantity in the applications mentioned, but also because solving the problem in the time domain can eventually help us in ascertaining the actual charging mechanisms and in finding methods for increasing the performance of pore-capacitance based techniques [1].

The key point rests on connecting the time and dimension scales of the macroporous and microporous regions in the electrode, illustrated in Fig. 2.1. We note that the focus will be placed on the charging/discharging of electrodes due exclusively to the building/expansion of electrical double layers (EDLs), without involving faradaic reactions, that is, without considering the possibility of electron transfer between the ions in solution and the electrode. This study is out of the scope of this work and has been analyzed in detail by Biesheuvel et al. [2].

The ideas to be developed in this chapter can be summarized as follows:

- Assume that we place a solution of given electrolyte concentration (typically NaCl, but multiionic solutions are also of great practical interest, as discussed in Chapter 5 of this book and also in [3, 4]) in contact with a pair of porous electrodes. We apply a certain potential difference between the electrodes by means of an external power source.
- We want to evaluate which are the final concentration and potential profiles throughout the electrodes and the spacer channel, and how long it takes to establish them.
- Furthermore, if the solution in the channel is suddenly changed, we are also interested in ascertaining the evolution of the quantities of interest.

Interface Science and Technology, Vol. 24. https://doi.org/10.1016/B978-0-12-811370-7.00002-4

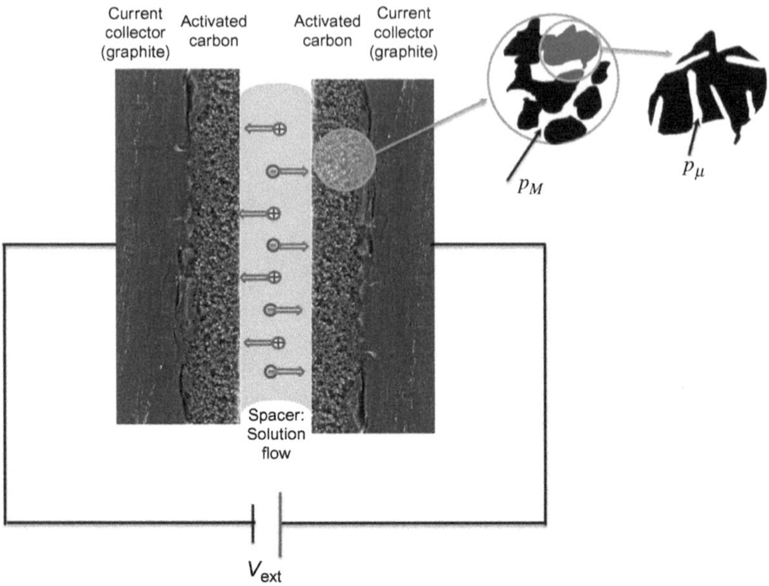

FIG. 2.1 The coexistence of macro- and micropores in the electrode volume. The respective porosities are p_M and p_μ.

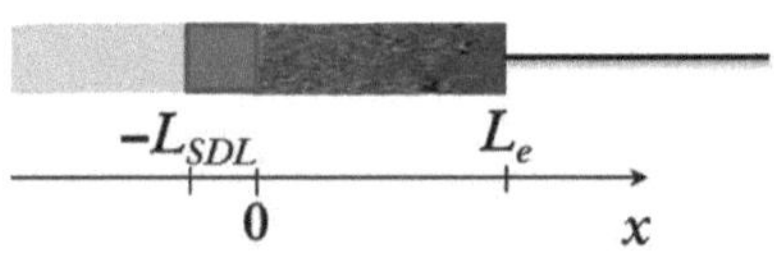

FIG. 2.2 From the channel to the current collector we distinguish three regions: bulk, stagnant diffusion layer, and active electrode.

- Eventually, we also wish to relate the observations with the charge-potential cycles in the CDLE technique, presented in Chapter 4.
- As another field where this study has relevance, the kinetics of capacitive deionization will also be dealt with.

2.2 Basic Physics of the Problem

Following the approach of Biesheuvel and Bazant [5, 6] and Rica et al. [7–9], we idealize the capacitive cell schematized in Fig. 2.1, as shown in Fig. 2.2: the spacer channel has a certain salt concentration c_{bulk} (we assume a monovalent electrolyte like NaCl; more complicated solution compositions are dealt with elsewhere [10, 11]). If we move from the center of the channel to the graphite collector, we will find three regions, namely, the channel or bulk volume, of homogenous properties; the *stagnant diffusion layer*, SDL (length L_{SDL}); and the active carbon electrode (dimension L_e). The SDL concept is important and stems from the fact that the transport of ions from a solution into a porous electrode involves the appearance of a region in the solution close to the electrode along which the concentration and potential change with distance [6].

Inside the electrode, the total porosity (p: relative total pore volume) is assumed to be the sum of macroporosity, p_M (average pore radius around or above 1 μm) and microporosity, p_μ (nanometer pore radius). The solution concentration in the bulk channel is c_{bulk}, and will change from that of sea to that of river water in a typical CapMix (capacitive mixing) process. Such a change will eventually be considered step-like. The origin of electric potential ($\Psi = 0$) will also be set at the spacer channel, which is equipotential and with constant concentration up to the contact with the SDL. Electrically, the latter will be considered as a purely resistive element, and the electric current density through it is considered constant, with value i_{SDL}.

When the external source is connected, establishing a potential difference between both electrodes, ions will move from or to the latter, subjected to the electric field and the concentration gradients. Only in the electrodes will charge and salt adsorption take place, and this will continue until the steady state is reached, characterized by the fact that the electrical current through the cell vanishes, so that the external potential is compensated for by the potential drop in the electrode, in turn, a superposition of that in the macropores plus that in the diffuse layer of the micropores and that in the Stern layer of the latter [7, 9]. After that, the concentration in the channel will be lowered (salt water out, fresh water in) in open circuit (ideally, constant charge on the electrodes), and a new steady-state situation will be reached. The final cell voltage must be determined, as well as the time needed to achieve it. A subsequent discharge through the external and internal resistances will bring the electrode to the power source potential again.

In the process of travelling along the channel, ions are adsorbed on their corresponding counter-electrodes, and desorbed from the electrodes with the same charge sign. Since the potential is typically high, adsorption dominates, and a net amount of salt is adsorbed on the electrodes. This mass balance must be also taken into account, although it is most significant in capacitive deionization, as described later in the chapter.

2.3 Equations and Boundary Conditions

The flux of ions of type i, j_i (mol/m^2 s), is determined by the Nernst-Planck equations:

$$j_i = -D\frac{dc_i}{dx} - \frac{ez_i c_i D}{k_B T}\frac{d\Psi}{dx}, \quad i = 1,2, \quad D_1 = D_2 = D, \quad z_1 = -z_2 = z \tag{2.1}$$

where a scalar form is used to account for the symmetry of the problem, whereby all quantities depend just on the distance along the electrode, x. Note that the concentrations are measured in mol/m^3, and that the sum

$j_{salt}=(j_1+j_2)$ measures the salt adsorption or desorption to/from the electrode, whereas the difference $i=zF(j_1-j_2)$ (F is the Faraday constant) is the electric current density. As mentioned, the salt concentration changes with time according to Fick's second law; in the case of the SDL, no salt adsorption takes place, and the equation reads:

$$\frac{\partial c_{SDL}}{\partial t}=D\frac{\partial^2 c_{SDL}}{\partial x^2} \tag{2.2}$$

In contrast, in the (thick EDLs of the) micropores salt adsorption/desorption takes place, and hence the concentration in the macropores must change accordingly:

$$\frac{\partial c_M}{\partial t}=D\frac{\partial^2 c_M}{\partial x^2}-\frac{p_\mu}{p_M}\frac{1}{2L_e}j_{salt} \tag{2.3}$$

The same can be said about electric current. In the SDL Ohm's law can be applied, with constant current density (F is the Faraday constant):

$$i_{SDL}=-\frac{2DzFc_{SDL}}{k_BT}\frac{\partial \Psi_{SDL}}{\partial x} \tag{2.4}$$

whereas in the macropores, the current will change with time again due to charge adsorption/desorption at the EDLs:

$$\begin{aligned} i_M &= -\frac{2DzFc_M}{k_BT}\frac{\partial \Psi_M}{\partial x} \\ \frac{\partial i_M}{\partial x} &= -\frac{p_\mu}{p_M}\frac{\partial \rho_\mu}{\partial t} \end{aligned} \tag{2.5}$$

where ρ_μ is the volume charge density in the micropores, to be given shortly. The boundary conditions are established as follows [7–9]:

- Continuity of the concentration at the interface electrode/SDL:

$$c_{SDL}=c_M=c_{bulk} \tag{2.6}$$

- Continuity of the electric potential at the same interface:

$$\Psi_{SDL}=\Psi_M=\Psi_0 \tag{2.7}$$

- Continuity of the electric current (modulated by the macroporosity in the macropore):

$$i_{SDL}=p_M i_M \tag{2.8}$$

- Continuity of salt flux (the ionic diffusion coefficient is assumed to be identical throughout the system):

$$\left.\frac{\partial c}{\partial x}\right|_{x=0^-}=\left.\frac{\partial c}{\partial x}\right|_{x=0^+} \tag{2.9}$$

- The electrodes are blocking for ions:

$$\left.\frac{\partial c_M}{\partial x}\right|_{x=L_e}=0$$
$$i_M\big|_{x=L_e}=0 \tag{2.10}$$

- Bulk values of the concentration and potential:

$$c\big|_{x=-L_{SDL}}=c_{bulk}$$
$$\Psi\big|_{x=-L_{SDL}}=0 \tag{2.11}$$

- Additionally, the current in the external circuit must equal that in the cell:

$$\frac{V_{cell}-V_{ext}}{R_{ext}}=i_{SDL}A_{elec} \tag{2.12}$$

where A_{elec} is the macroscopic area of the electrodes. The cell voltage V_{cell} (potential difference between the elecrodes) can be calculated in terms of the diffuse and Stern potential drops of the EDLs in the micropores (see Chapter 1) plus the voltage drop in the macropores, minus that in the internal resistance of the cell:

$$V_{cell}=2(\Delta\Psi_M+\Delta\Psi_d+\Delta\Psi_{Stern})_{x=0}-\Delta V_{int}$$
$$\Delta V_{int}=I_{ext}R_{int} \tag{2.13}$$

where the main contribution to the internal resistance comes from electrolyte in the spacer channel, being thus much smaller in the presence of salt water than in fresh water.

The surface charge density associated with the macropore thin double layers can be written (Chapter 1 and Ref. [12]):

$$\sigma_M=-\frac{2k_BT\varepsilon\varepsilon_0}{ze\kappa^{-1}}\sinh\left(\frac{ze\Delta\Psi_d}{2k_BT}\right) \tag{2.14}$$

This potential drop in the uncharged Stern layer can also be calculated in terms of this surface charge:

$$\Delta\Psi_{Stern}=-\frac{\sigma_M}{C_{St}}=-\frac{\sigma_M\delta_\mu}{\varepsilon\varepsilon_0} \tag{2.15}$$

with δ_μ the thickness of the Stern layer on the micropores. The salt adsorption (mol/m^2) on these interfaces is:

$$\Gamma_M^{excess}=\int_d^\infty\left(c_Me^{-\frac{ez\Psi}{k_BT}}+c_Me^{\frac{ez\Psi}{k_BT}}-2c_M\right)dx=$$
$$=4\frac{c_M}{\kappa}\left[\cosh\left(\frac{ze\Delta\Psi_D}{2k_BT}\right)-1\right]=8\frac{c_M}{\kappa}\sinh^2\left(\frac{ze\Delta\Psi_D}{4k_BT}\right) \tag{2.16}$$

Let us mention finally that the modified Donnan model (see Chapter 1 and Refs. [2, 6, 13]) is used for characterizing the micropores. This means that the potential is assumed to be constant and equal to $\Delta\Psi_d$ throughout the whole micropore section, and so is the ionic concentration $c_{i,\mu}$:

$$c_{i,\mu} = c_{i,M}\exp\left(-z_i e\Psi_D/k_BT\right) \tag{2.17}$$

The volume charge density in the micropores is hence:

$$\rho_\mu = -c_M F \sinh\left(\frac{ez\Delta\Psi_d}{k_BT}\right) = -\Delta\Psi_{Stern}C_{Stern}^{vol}$$
$$C_{Stern}^{vol} = C_{Stern}S_{pore}/V_{pore} \tag{2.18}$$

and the salt adsorption:

$$\Gamma_\mu^{excess} = \frac{1}{2}(c_1 + c_2)_\mu - c_M = 2c_M\sinh^2\left(\frac{ze\Delta\Psi_D}{2k_BT}\right) \tag{2.19}$$

2.4 Numerical Solution: Concentration and Potential Profiles With Time

The different time and dimension scales of the problem suggest that equations be solved in a dimensionless fashion. We denote all dimensionless quantities by a tilde. The substitutions carried out are detailed in Table 2.1.

With these substitutions, Biesheuvel and Bazant [6] and Rica et al. [7] demonstrated that the system is reduced to a pair of partial differential equations for the salt concentration $\tilde{c}_M$ and adsorption as follows:

TABLE 2.1 Substitutions Employed in Nondimensionalizing the Quantities

Quantity	Dimensionless Form
Distance	$\tilde{x} = x/L_e$
Potential	$\left(\tilde{\Psi}, \tilde{V}\right) = (\Psi, V)/(k_BT/ze)$
Concentration	$\tilde{c}_i = c_i/c_{bulk}$
Ion flux	$\tilde{j}_i = j_i/(2Dc_{bulk}/L_e)$
Time	$\tilde{t} = t/\left(L_e^2/D\right)$
Electric current density	$\tilde{i} = i/(2c_{bulk}zeD/L_e)$
Surface charge density	$\tilde{\sigma} = \sigma/\left(2e\kappa^{-1}c_{bulk}\right)$
Adsorbed salt	$\tilde{\Gamma}_M^{excess} = \frac{\kappa}{2\sqrt{c_M c_{bulk}}}\Gamma_M^{excess}$; $\tilde{\Gamma}_\mu^{excess} = \frac{\Gamma_\mu^{excess}}{c_M}$

$$\frac{\partial}{\partial \tilde{t}}\left(p_M \tilde{c}_M + \gamma \Gamma^{excess}_{M,\mu}\right) = p_M \frac{\partial^2 \tilde{c}_M}{\partial \tilde{x}^2}$$
$$\gamma \frac{\partial \tilde{\sigma}}{\partial \tilde{t}} = p_M \frac{\partial}{\partial \tilde{x}}\left(\tilde{c}_M \frac{\partial \tilde{\Psi}}{\partial \tilde{x}}\right) \tag{2.20}$$

where the coefficient γ equals p_μ in the micropores and $\kappa^{-1}S_{pore}/V_{pore}$ in macropores. The meaning of Eqs. (2.20) is [6]:

- The first one indicates the equality between the salt adsorbed in the micropores and the flux of salt in the macropore solution.
- The second one establishes the equality of the charging of the EDLs and the divergence of the ionic current.

Solving them numerically, it is possible to find the potential and concentration profiles along the electrodes, which behave symmetrically. A detailed analysis is provided in [7], and Biesheuvel and Bazant [6] offer results in the simple case that the Stern layer is absent, but which can qualitatively help us in understanding how the potential and the concentration evolve with time (Fig. 2.3).

Note how the potential at the electrode, initially equal to half the applied voltage difference, decreases with time as the electrode gets charged and the diffuse potential increases, thus reducing the difference between V_{cell} and V_{ext}. At the same time, the salt content of the macropores tends to decrease as a consequence of ion adsorption in the EDLs of the micropores (counterions are adsorbed and coions are desorbed).

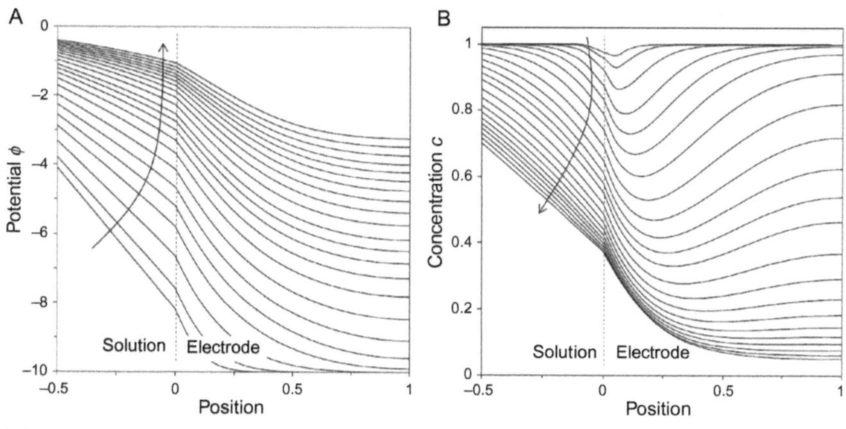

FIG. 2.3 Dimensionless potential (A) and salt concentration (B) in a porous electrode after application of a potential step (dimensionless value −10; $V_{ext} = -20 \times k_BT/e = -51$ mV). The arrows indicate the direction of time increase. The SDL and electrode lengths are equal to 100 μm. The bulk electrolyte concentration is 10 mM NaCl, the porosity is 0.5, and $S_{pore}/V_{pore} = 2 \times 10^7 \text{m}^{-1}$. *Reproduced from Biesheuvel PM, Bazant MZ: Nonlinear dynamics of capacitive charging and desalination by porous electrodes, Phys Rev E 2010;81:031502. This article or its components as it is available under the terms of the Creative Commons Attribution 3.0 License.*

2.5 Implications for Energy Production

Having solved the kinetics of salt and charge adsorption, it is possible to simulate a CDLE energy harvesting cycle (Chapter 4):

1. The cell in Fig. 2.1 is filled with salt water (500 mM NaCl) and contacted (typically for several hours) with the external power source (V_{ext} applied).
2. Fresh water solution (20 mM) is pumped in open circuit (constant charge). The potential difference between the electrodes, V_{cell}, is recorded.
3. The cell is connected again with the external power source through an external resistor: current flows while the electrodes discharge.
4. Salt water is flowed again in open circuit.
5. The circuit is closed and the electrodes are charged in the presence of the salt solution, starting the cycle again.

Fig. 2.4 shows actual results with one of the cells used in our laboratory (see details in Chapter 6), and Fig. 2.5 includes the theoretical predictions of the model developed in [9] for a similar cell. Note how the model effectively captures the main features of the process, using reasonable values for the dimensions of both the electrode and the SDL, and of the Stern volume capacitance:

- The voltage rise associated with the passage of fresh water because of the reduction in EDL capacitance. The experimental data are typically around some tens of mV

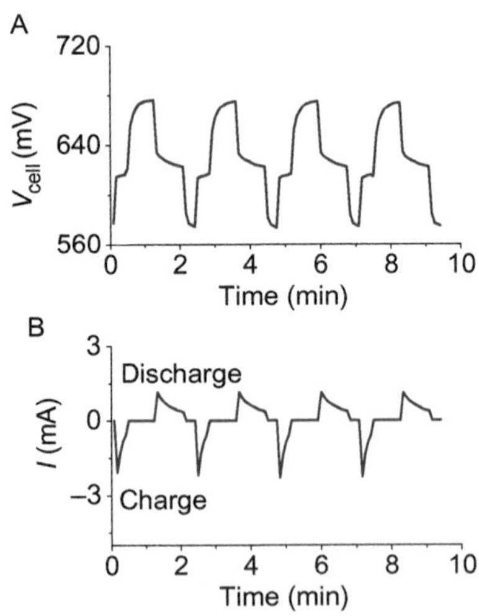

FIG. 2.4 Cell potential (A) and current (B) in a CDLE experiment carried out with one of the cells described in Chapter 6, with circular electrodes 1 cm in radius. The external resistance was 10 Ω in fresh water and 2 Ω in salt water; the solutions exchanged were NaCl 600 mM and 20 mM.

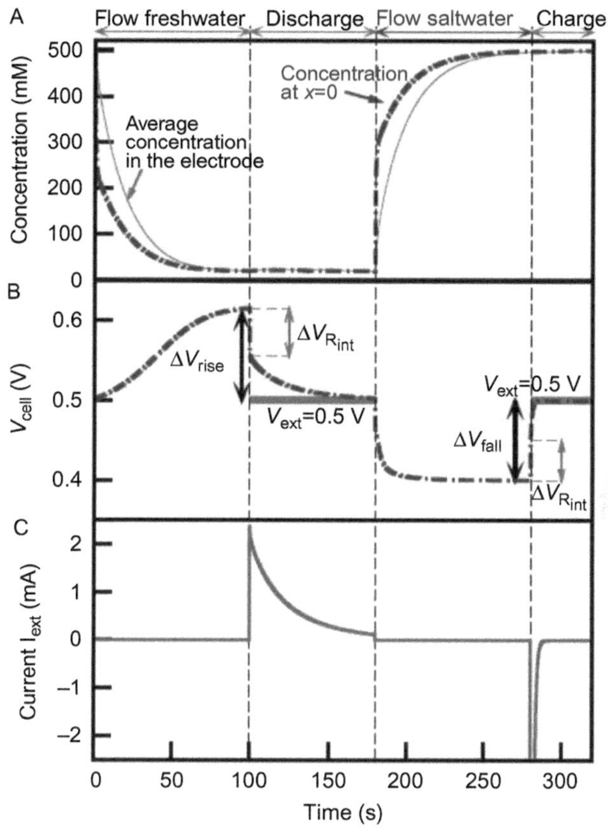

FIG. 2.5 Calculations of the time evolution of salt concentration in the micropores (top), cell voltage (middle), and current in the external circuit (bottom) during the CDLE cycle. Parameters: electrode area 2.25 cm^2; electrode separation: 1 mm; solutions exchanged: 500 mM and 20 mM NaCl; SDL and electrode lengths: 150 μm; $C_{Stern}^{vol}=0.1$ GF/m^3; porosities: $p_M=p_\mu=0.3$; external resistors: 25 Ω (fresh solution), 1 Ω (salt water). *Reproduced with permission from Rica RA, et al.: Electro-diffusion of ions in porous electrodes for capacitive extraction of renewable energy from salinity differences, Electrochim Acta 2013;92:304–314. Copyright 2013 Elsevier Ltd.*

- The voltage drop in the internal resistance, when connecting to the external circuit for discharging
- The few milliamps current for the cell dimensions simulated
- The voltage drop when the salt water is again pumped in the channel
- The different time scales for the two main events, rise in fresh water, drop in salt water
- In addition, it describes the evolution of the salt concentration in the macropores: the EDLs release salt when contacted with the fresh solution and, vice versa, accumulate it in the presence of the salty water.

2.6 Capacitive Deionization Cycles

The theory of capacitive deionization [2, 6, 14] is focused on the concentration changes in the channel when a potential difference between both electrodes is established. The ions will migrate from one electrode to the other until steady state is reached and salt adsorption has taken place. Under constant voltage conditions, the electrical current through the cell will reach a maximum value and after that will vanish when the external potential is compensated for by the potential drop in the electrodes, that is, the macropores, the diffuse layer of the micropores, and the Stern layer of the latter. The aim is to predict the concentrations at the outlet of the cell as a function of time.

The 1-D model (only one spatial variable along the flow through the cell) that we are going to use consists of dividing the cell into a certain number M of subcells perpendicular to the direction of the flow through the spacer channel Fig. 2.6, in such a way that each of them will be at a given concentration $c_{sp,j}$, where j represents the subcell number. The spacer is in close contact with the macropores, which constitute the pathway to the micropores, so that it can be assumed that the concentration at the macropores is the spacer concentration $c_{M,j} \equiv c_{sp,j}$. Thus the concentration in the first subcell, $j=1$, is that of the solution to be desalinated, $c_{M,1}=c_{IN}$, and that of the last subcell is the output concentration $c_{M,M}=c_{OUT}$.

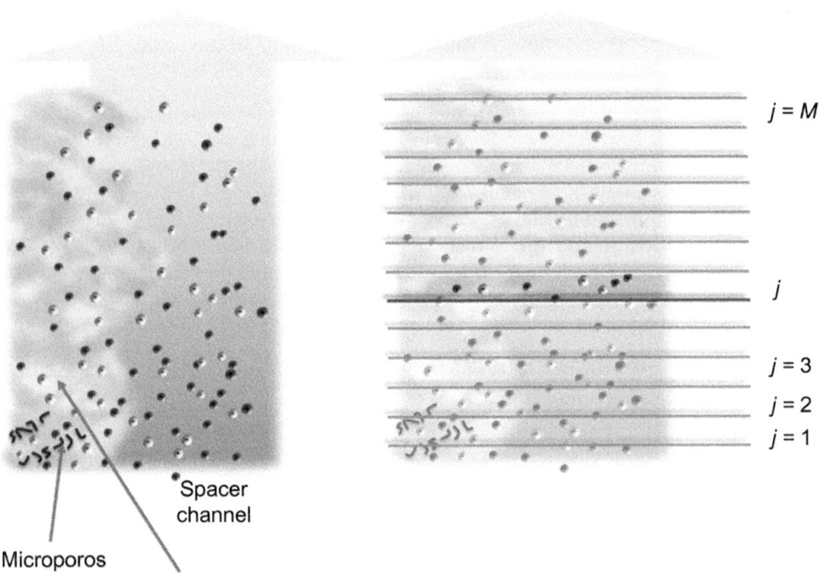

FIG. 2.6 Left: sketch of half-cell including the electrodes with macropores and micropores in contact with the spacer channel. Right: the j-th subcell, where $j=1,\ldots,M$.

For each of the regions of the cell, we will write the mass balance equation. In the case of the spacer channel, the accumulation of ions at a given subcell j is determined by the diffusive and migratory flux of ions of type i, $j_{i,j}$, in the x direction from macropores to micropores (perpendicular to the flow) and by the convective transport of ions from the previous subcell.

$$L_{sp}\frac{dc_{i,M,j}}{dt} = -j_{i,j} + \frac{M\phi_v}{A_{sub}}\left(c_{i,M,j-1} - c_{i,M,j}\right) \tag{2.21}$$

where L_{sp} is the length of the spacer channel, A_{sub} the projected area of the electrodes, and ϕ_v the pumping speed of the solution through the cell. A simplified model considers that only electromigration takes place and the Nernst-Planck equation (Eq. 2.1) can be simplified, neglecting the gradient of salt along the macropores (in the direction perpendicular to the flow of solution) [11, 15–17]:

$$j_{i,j} = z_i D_i c_{i,M,j} L_{eff}^{-1} \Delta\phi_{tr,j} \tag{2.22}$$

where L_{eff} is an effective length determining the average electric field between electrodes, and $\Delta\phi_{tr,j}$ is the potential difference that originates the transport of ions to micropores, similar to that defined in Eq. (2.13) although in this model is reduced to:

$$\Delta\phi_{tr,i} = V_{cell} - 2|\Delta\psi_{Stern} + \Delta\psi_d| \tag{2.23}$$

which is half the applied cell voltage minus the voltage drop in the EDLs of the electrode. In a more elaborate model, the potential decay in the spacer and the electrodes themselves can also be taken into account, and it will depend on the resistance of both regions, namely, the spacer channel, and the electrode [18], where it is necessary to appeal to the porous electrode model to characterize this contribution. This expression agrees with Eq. (2.18) relating the charge density with the potential difference in the Stern and diffuse layers. However, an improvement of the modified Donnan model described in the previous section is taken into account to include a chemical attraction energy for ion adsorption inside the carbon particles [10, 19]. This nonelectrostatic attraction was experimentally justified by the fact that ions can penetrate inside uncharged pores. It can be introduced in the model by an additional term, μ_{att}, in the Boltzmann distribution of ions (2.17):

$$c_{i,\mu} = c_{i,M}\exp\left(-z_i e\Psi_D/k_B T + \mu_{att}\right) \tag{2.24}$$

This term can be different for each ionic species. The differential equations (2.21) and (2.25) can be solved to predict the concentration of ions at the exit of the cell and the charge density accumulated at the micropores. The predictions are illustrated in Fig. 2.7 showing the excellent agreement with the experimental data at different cell voltages and feed ionic concentrations.

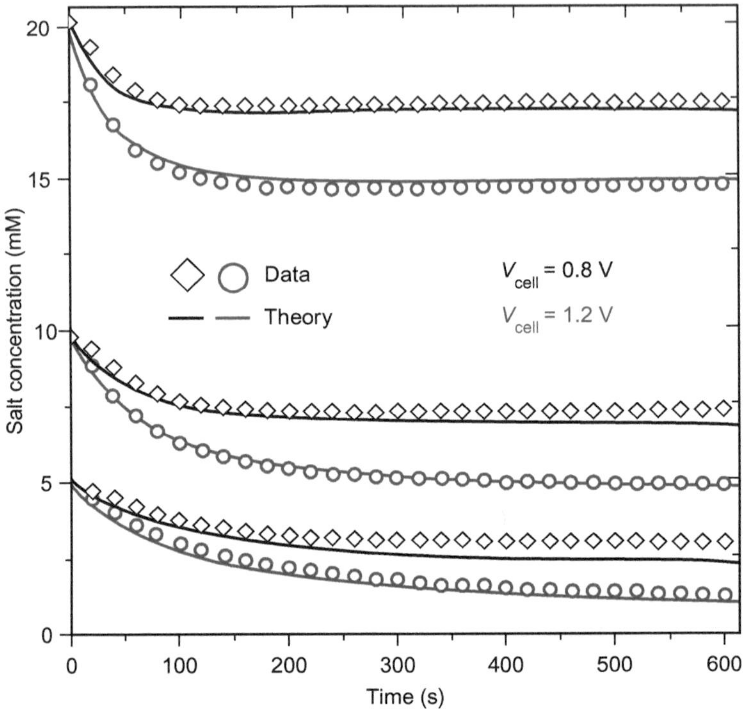

FIG. 2.7 NaCl concentration as a function of time in batch mode regime for three values of ionic concentration and two applied cell voltages. Symbols: experimental data; Lines: theoretical predictions. Stack of eight cells with $A_{sub} = 33.8\,cm^2$, $L_{sp} = 250\,\mu m$, $C_{St} = 0.21\,GF/m^3$ and $\phi_v = 60\,mL/min$ for the stack. *Reproduced with permission from Porada S, et al.: Review on the science and technology of water desalination by capacitive deionization, Prog Mater Sci 2013;58:1388–1442. Copyright 2013 Elsevier Ltd.*

If ions are removed from the solution and transported to micropores, the charge density in the micropores $\rho_{\mu,j}$ in contact with macropores in subcell j should change:

$$-L_e \frac{d}{dt}\left(p_\mu \rho_{\mu,j}\right) = \sum_{i=1}^{N} z_i J_{i,j} \tag{2.25}$$

It is worth noting that the ionic concentration in Eq. (2.22) is that at the macropores, a condition that will change if a treated carbon should be modeled. In the particular case of polyelectrolyte-coated or *soft* electrodes, the ionic fluxes from macropores to micropores are modified by the coating layer, favoring the counterions entrance and restraining the exit of coions, as schematized in Fig. 2.8.

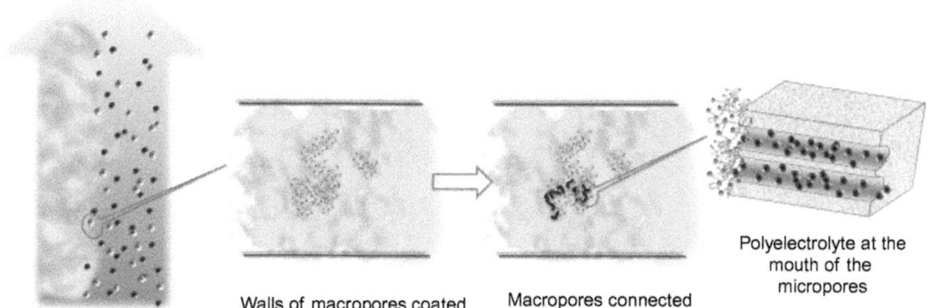

FIG. 2.8 Sketch of CDI cell in a soft electrode assembly. *Reproduced with permission from Ahualli S, et al.: Use of soft electrodes in capacitive deionization of solutions, Environ Sci Technol 2017;51:5326–5333. Copyright 2017 American Chemical Society.*

The model must take into account the presence of the polyelectrolyte layer in the flux calculations and replacing the macropore concentration in Eq. (2.22) by the soft layer concentration that would be in contact with the micropores:

$$J_{i,j} = z_i D_i c_{i,M,j}^{SE} L_{eff}^{-1} \Delta\phi_{tr,j} \quad (2.26)$$

where the concentration at the soft layer $c_{i,M,j}^{SE}$ can be calculated as described in Chapter 7. The blocking of coions improves the salt adsorption and increases the efficiency, as discussed later.

2.7 Membrane Capacitive Deionization Cycles

With regard to the model of membrane capacitive deionization (MCDI), the main difference with respect to the CDI theory just described is that the macropore concentration is no longer the same as the spacer ionic concentration. The IEM is located between them and it will control the ionic flux, concentrations, potential decay, and, of course, the dynamics of the process. The sketch of this MCDI assembly is represented in Fig. 2.9, where the cell is also subdivided into several subcells for the 1-D model used.

A single salt concentration corresponds to each subcell neglecting a salt concentration gradient in the x direction. This is different from the approach in [20] and the one described in the previous section, where the concept of a stagnant boundary layer close to either the electrode or to the membrane was considered. In this model [21] it is assumed that the spacer channel is quite thin, and thus a theoretical separation of the channel into a bulk-phase and two diffusion films is not possible. Furthermore, the model assumes that ion transport resistance is mainly located within the electrode, so that concentration variations across the spacer channel are insignificant.

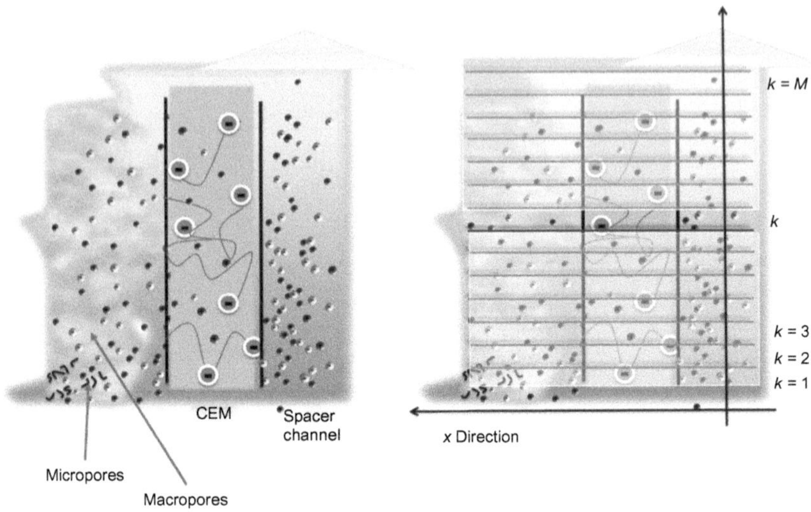

FIG. 2.9 Sketch of half cell: One porous electrode in contact with a cation exchange membrane and the spacer channel.

The mass balance equation will be written for each region of the cell, namely spacer, ion exchange membrane, macropores, and micropores. It is worth stressing that while in the CDI model the governing equations were written for each ionic species, here we only deal with total salt concentration, because charge neutrality is assumed at both sides of the membranes, that is, spacer and macropores. Therefore, the ion flux, j, refers to the total contribution of both types of ions.

In the case of the spacer channel where the solution flows, the accumulation of salt is determined by the diffusive and electromigratory flows through the membrane, j, in the x direction (perpendicular to the flow) and by the convective transport of ions from the previous subcell. Writing this balance for subcell k we find:

$$L_{sp}\frac{dc_{sp,k}}{dt} = -j_k + \frac{M\phi_v}{A_{sub}}\left(c_{sp,k-1} - c_{sp,k}\right) \tag{2.27}$$

The ionic flux is determined by the properties of the membrane, its thickness L_{mem}, its charge, ρ_{mem}, and the diffusion coefficient of the ions within it. For highly charged membranes, the Nernst-Planck equation can be linearized and the flow through it takes the following form:

$$j_k = -\frac{D}{L_{mem}}\left[\Delta c_{T,mem,k} - \rho_{mem}\Delta\psi_{mem,k}\right] \tag{2.28}$$

where the total concentration of ions inside the membrane, $c_{T,mem}$, is, assuming Donnan equilibrium inside it, given by:

$$c_{T,mem,k} = \sqrt{\rho_{mem}^2 + (2c_{salt,k})^2} \tag{2.29}$$

where c_{salt} will take the value c_{sp} at the membrane-spacer channel interface, and c_M at the membrane-macropore interface. Then $\Delta c_{T,mem,k}$ is the difference between both limiting values. The potential difference of the membrane is linearly related to the current as:

$$i_k = -\frac{D}{L_{mem}}\langle c_{T,mem,k}\rangle \Delta\psi_{mem,k} \tag{2.30}$$

In this linearized model of the membrane it is assumed that the flow of ions is constant across the membrane and therefore there is no salt accumulation. In addition, nonideal membranes are considered so that the ion current is mainly due to counterion transport, but it also has a coion contribution in the reverse direction.

The next step is the balance of salt in the electrode. The flow of ions that go through the membrane changes the macropore concentration and feeds the micropores, altering the concentration inside them, according to Eq. (2.3):

$$\frac{d}{dt}\left(2p_M c_{M,k} + p_\mu c_{\mu,k}\right) = \frac{j_k}{L_e} \tag{2.31}$$

The relationship between concentrations in macro- and micropores are given by the modified-Donnan model (2.17).

As the macropore is electrically neutral, the excess of charge in the micropores is determined by the current i_k:

$$L_e \frac{d}{dt}\left(p_\mu \rho_{\mu,k}\right) = i_k \tag{2.32}$$

Two boundary conditions must be fulfilled. One refers to current continuity for each subcell: the current is invariant in the x direction, that is, the current across the spacer (2.33), the membranes, and its interfaces (2.30) and that going into the micropores (2.32) are all identical. In the spacer channel, a constant resistance is assumed in such a way that the current is proportional to the potential decay:

$$i_k = 2Dc_{sp,k}\frac{\Delta\psi_{sp,k}}{L_{sp}} \tag{2.33}$$

where D is an average between the diffusion coefficients of both ionic species. The intensity I results from integration of i_k over the electrode area multiplied by the Faraday number, F.

The other boundary condition is that half the external voltage applied between electrodes is the superposition of potential drop over each spacer channel, $\Delta\psi_{sp}$, the Donnan potential at the membrane/spacer edge, $\Delta\psi_{d,memb/sp}$, the potential across the membrane, $\Delta\psi_{memb}$, minus Donnan potential at the membrane/electrode edge, $\Delta\psi_{d,memb/elect}$, the potential drop in the electrode, $\Delta\psi_{elect}$, and the potential decay at the micropores, including the Donnan potential and the Stern layer potential, $\Delta\psi_d+\Delta\psi_{St}$:

$$\frac{1}{2}V_{cell} = \Delta\psi_{sp} + \Delta\psi_{d,memb/sp} + \Delta\psi_{memb} - \Delta\psi_{d,memb/elec} + \Delta\psi_{elec} + (\Delta\psi_d + \Delta\psi_{St})_{mi} \tag{2.34}$$

At the two membrane/solution interfaces we have Donnan equilibrium so that the potential at each edge is given by:

$$\Delta\psi_{d,memb/sp} = \sinh^{-1}\frac{\rho_{mem}}{2c_{sp}} \tag{2.35}$$

$$\Delta\psi_{d,memb/elec} = \sinh^{-1}\frac{\rho_{mem}}{2c_M} \tag{2.36}$$

Within the porous electrode, the full nonlinear porous electrode theory for desalination, described in [6] and in Section 2.3, is not used. A simplified approach is considered in which the electrode region is characterized by an electrical resistance linearly dependent on macropore salt concentration, which is assumed spatially invariant inside it (no salt or potential gradients):

$$\Delta\psi_{elect} = -\frac{IFR_{elect}}{c_M} \tag{2.37}$$

with R_{elec} an electrode specific resistance. It is an equation similar to Eq. (2.33) for the spacer channel.

The predictions of the model regarding macropore salt concentration are displayed in Fig. 2.10. Note the essential difference between CDI and MCDI: in the former the macropore and spacer regions are essentially indistinguishable. For every unit electronic charge, either counterion adsorption (removed from the solution in the channel) or coion desorption (ending in the solution in the spacer) will take place. As a result, the concentration in the macropore will follow the same trend as in the spacer, namely, decrease until the micropores are saturated and subsequently increase towards the feed solution concentration. In contrast, in MCDI, macropores and spacer are separated by the membrane, and while the concentration in the latter will evolve similar to CDI, this will not occur in the macropores. Here the concentration increases by desorption of coions from the micropore being neutralized by counterions transported through the membrane from the spacer. This process favors the overall salt adsorption, now due to the contributions of both micro- and macropores. Additionally, another consequence can be mentioned: since coions are not transported into the spacer channel, the overall charge efficiency of the process approaches unity (one ion removed for each electron transported). Comparison with experimental data is very favorable, as shown in Fig. 2.11, so the main features of the MCDI process are in fact captured.

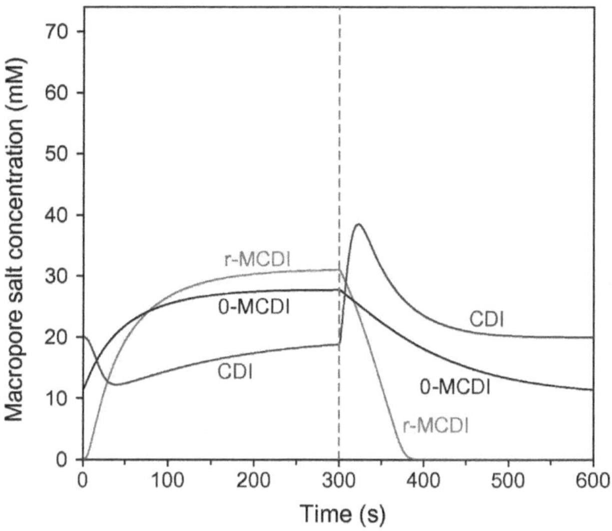

FIG. 2.10 Theoretical predictions of macropore salt concentration using CDI and MCDI under ZV (0-MCDI) and RV (r-MCDI). *Reproduced with permission from Biesheuvel M, et al.: Theory of membrane capacitive deionization including the effect of the electrode pore space, J Colloid Interface Sci 360:239–248, 2011. Copyright 2011 Elsevier Ltd.*

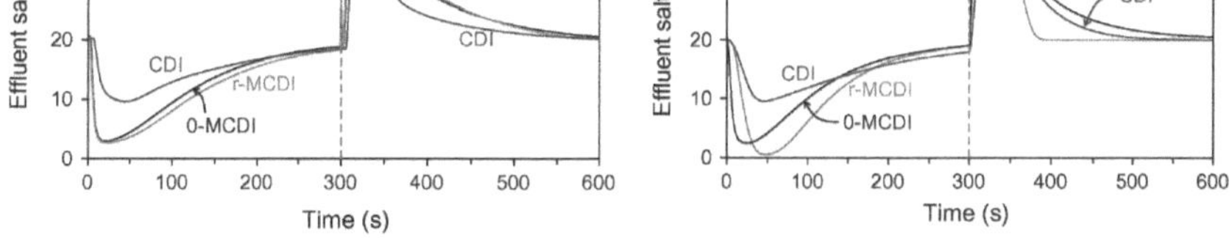

FIG. 2.11 Effluent NaCl concentration using the same conditions of Fig. 2.10. $V_{cell} = 1.2\,V$ and $c_{IN} = 20\,mM$. *Reproduced with permission from Biesheuvel M, et al.: Theory of membrane capacitive deionization including the effect of the electrode pore space, J Colloid Interface Sci 2011;360:239–248. Copyright 2011 Elsevier Ltd.*

2.8 Conclusions

In this chapter we have gone into some details on the kinetics of the processes of ions and salt adsorption in the micropores of active electrodes. This analysis is important from two points of view: understanding the phenomena involved and using this knowledge for explaining experimental results and, even more important, predicting improvements in the implementation. Consideration is first paid to the case of energy harvesting by EDL expansion (CDLE). A 1-D model is described that allows predicting the time evolution of charge and potential inside the macro- and micropores after application of a step electric field. The same model yields the potential rises and drops when salty and fresh solutions are exchanged. The same treatment can be applied, with slight modifications, to the evaluation of the kinetics of capacitive deionization, which can be considered the long-time limit of the solutions found for the first problem. The cases of treated (soft electrode CDI) and membrane-coated electrodes are also considered. An excellent agreement with experimental data is reported, indicating that the physics of the CDI processes is in fact captured.

Acknowledgments

Financial support for this work from Junta de Andalucía, Spain (Project PE2012-FQM-0694), and University of Granada ("Vicerrectorado de Investigación, Plan Incorporación Jóvenes Doctores") is gratefully acknowledged.

References

[1] Rossi M, Wallmersperger T, Neukamm S, Padberg-Gehle K. Modeling and simulation of electrochemical cells under applied voltage. Electrochim Acta 2017;258:241–54.

[2] Biesheuvel PM, Fu Y, Bazant MZ. Diffuse charge and Faradaic reactions in porous electrodes. Phys Rev E 2011;83(6):061507.

[3] Fernandez MM, Ahualli S, Iglesias GR, Gonzalez-Caballero F, Delgado AV, Jimenez ML. Multi-ionic effects on energy production based on double layer expansion by salinity exchange. J Colloid Interface Sci 2015;446:335–44.

[4] Jimenez ML, Ahualli S, Arenas-Guerrero P, Fernandez MM, Iglesias G, Delgado AV. Multiionic effects on the capacitance of porous electrodes. Phys Chem Chem Phys 2018;20(7):5012–20.

[5] Bazant MZ, Thornton K, Ajdari A. Diffuse-charge dynamics in electrochemical systems. Phys. Rev. E 2004;70(2):021506.

[6] Biesheuvel PM, Bazant MZ. Nonlinear dynamics of capacitive charging and desalination by porous electrodes. Phys. Rev. E 2010;81(3):031502.

[7] Rica RA, Brogioli D, Ziano R, Salerno D, Mantegazza F. Ions transport and adsorption mechanisms in porous electrodes during capacitive-mixing double layer expansion (CDLE). J Phys Chem C 2012;116(32):16934–8.

[8] Brogioli D, Ziano R, Rica RA, Salerno D, Mantegazza F. Capacitive mixing for the extraction of energy from salinity differences: survey of experimental results and electrochemical models. J Colloid Interface Sci 2013;407:457–66.

[9] Rica RA, Ziano R, Salerno D, Mantegazza F, Bazant MZ, Brogioli D. Electro-diffusion of ions in porous electrodes for capacitive extraction of renewable energy from salinity differences. Electrochim Acta 2013;92:304–14.

[10] Biesheuvel PM, Fu Y, Bazant MZ. Electrochemistry and capacitive charging of porous electrodes in asymmetric multicomponent electrolytes. Russ J Electrochem 2012;48(6): 580–92.

[11] Zhao R, van Soestbergen M, Rijnaarts HHM, van der Wal A, Bazant MZ, Biesheuvel PM. Time-dependent ion selectivity in capacitive charging of porous electrodes. J Colloid Interface Sci 2012;384:38–44.

[12] Hunter RJ. Foundations of colloid science. Oxford: Clarendon Press; 1987.

[13] Wang L, Biesheuvel PM, Lin SH. Reversible thermodynamic cycle analysis for capacitive deionization with modified Donnan model. J Colloid Interface Sci 2018;512:522–8.

[14] Johnson AM, Newman J. Desalting by means of porous carbon electrodes. J Electrochem Soc 1971;118(3):510–7.

[15] Tang W, Kovalsky P, Cao B, He D, Waite TD. Fluoride removal from brackish groundwaters by constant current capacitive deionization (CDI). Environ Sci Technol 2016;50(19):10570–9.

[16] Kim T, Dykstra JE, Porada S, van der Wal A, Yoon J, Biesheuvel PM. Enhanced charge efficiency and reduced energy use in capacitive deionization by increasing the discharge voltage. J Colloid Interface Sci 2015;446:317–26.

[17] Porada S, Borchardt L, Oschatz M, Bryjak M, Atchison JS, Keesman KJ, Kaskel S, Biesheuvel PM, Presser V. Direct prediction of the desalination performance of porous carbon electrodes for capacitive deionization. Energy Environ Sci 2013;6(12):3700–12.

[18] Zhao R, Satpradit O, Rijnaarts HHM, Biesheuvel PM, van der Wal A. Optimization of salt adsorption rate in membrane capacitive deionization. Water Res 2013;47(5):1941–52.

[19] Porada S, Bryjak M, van der Wal A, Biesheuvel PM. Effect of electrode thickness variation on operation of capacitive deionization. Electrochim Acta 2012;75:148–56.

[20] Biesheuvel PM, van der Wal A. Membrane capacitive deionization. J Membr Sci 2010;346(2):256–62.

[21] Biesheuvel PM, Zhao R, Porada S, van der Wal A. Theory of membrane capacitive deionization including the effect of the electrode pore space. J Colloid Interface Sci 2011;360(1): 239–48.

Chapter 3

The Electrical Double Layer as a Capacitor. Evaluation of Capacitance in Different Solutions: Effect of Ion Concentrations, Sizes, and Valencies

María L. Jiménez, Silvia Ahualli and María M. Fernández
Department of Applied Physics, University of Granada, Granada, Spain

3.1 Introduction

In recent years research in energy storage has been boosted by the appearance of double-layer capacitors [1, 2], basically consisting of a pair of highly porous electrodes in contact with an ionic solution. If there is a potential difference between them, ions in the solution form the electric double layer (EDL) simply by diffusion and electrostatic attraction, producing a huge amount of stored charge in the electrodes. The EDL special properties explain the name of "supercapacitors" given to these devices, with the main advantage of their rapid and reversible ionic adsorption mechanisms. Furthermore, they can be used not only for energy and charge storage, but it has been also demonstrated that such technology can be used to produce energy [3–6] or desalinate salty water [7–10].

The use of activated carbon particles as electrode material has the advantage of providing a huge amount of surface area by decreasing the pore size to nanometer scale. However, despite the fact that the traditional material has been activated carbon particles [1], recent investigations have shown that considerable improvement is possible. For instance, in Ref. [11], it is shown that not all the available surface participates in capacitive energy extraction based on Double Layer Expansion (CDLE). Thus it has been found [12, 13] that hierarchical porous material, in which small micropores ensure a large BET area and larger mesopores favor the access of ions to the small ones,

Interface Science and Technology, Vol. 24. https://doi.org/10.1016/B978-0-12-811370-7.00003-6

improves the kinetics of ionic transport. Other materials such as nanotubes [14] or holey graphene [15] have been also proposed.

All these technologies have two common points. (i) Ions are subjected to extreme confinement conditions: the size of micropores is about 2 nm or lower, which is just one order of magnitude larger than the hydrated ionic size. In these conditions, EDL overlap is likely, and hence its structure differs from that close to a surface in an unbounded medium. (ii) The electrodes must operate with moderate voltages and hence steric effects are guaranteed. The structure of the EDL, therefore its capacitance, is greatly affected by ion-ion interactions. Even for the most simple situation, with voltages as low as 0.1 V and the standard 20 mM dilute concentration of fresh water, neglecting steric repulsion gives unrealistic ionic concentrations in the proximity of the electrode [16]. Accordingly, major efforts have been made in order to analyze the steric repulsion contribution [16–20].

The need for a realistic description of the structure of the electrode/solution interface is more evident when these processes are tested with multiionic solutions. For example, it is well known that in the process of removing monovalent ions from a solution in the so-called capacitive deionization techniques (CDI), the presence of divalent ions produces the opposite effect [21–23]. In fact, in a theoretical and experimental investigation of the kinetics of Na^+ and Ca^{2+} adsorption in the EDL of porous and flat electrodes, Zhao et al. [24] found that the monovalent cations are gradually replaced by the divalent ones in the EDL, and they are also the first ones to be released when the CDI cell is short-circuited. The larger affinity of these ions for the surface was analyzed through a mean field theory in [25], including the nonideal ionic behavior in the model by using the excess potential of the Bikerman equation. This work focused on the CDLE technique and succeeded in explaining the experimental results: the presence of tiny amounts of divalent ions, despite increasing the surface charge, works against energy production.

One alternative to the Bikerman approach is the Carnahan-Starling (C-S) equation [26]. This has been demonstrated to provide a more accurate representation of the equation of state for a gas of rigid spheres than the Bikerman equation [19, 27–29], and has been successfully applied to reproduce the experimental capacitance values for the EDL without confinement restrictions [30]. In this work, the differences between the C-S model and Bikerman equation are analyzed. In the case of multiionic solutions, a realistic description of the kind of ions in the EDL can be of utmost importance in some applications, such as CDI applied to selective removal of one particular kind of ion by its adsorption in the EDL [24, 31]. In these cases, the equation of state must take into account the differences between size and valency of the ions, as the equation of state of Boublik-Mansoori-Carnahan-Starling-Leland (BMCSL) does [17, 32, 33].

Although none of these models was originally elaborated with focus on the nanoscopic geometry of the micropores in supercapacitors, CDLE, and CDI techniques, it is expected that the use of C-S or BMCSL will provide accurate results in the evaluation of the EDL capacitance in confined media.

A detailed description of the EDL with a local-density model has been given in [20]. In this work, the authors compare the C-S local-density model and molecular dynamics simulations. A detailed analysis of the structure of the EDL reveals free charge oscillations due to the strong steric ion correlations when they are densely packed. These oscillations cannot be predicted by a mean field analysis. Despite this limitation, the authors demonstrate an excellent agreement between the capacitances obtained with both approaches.

3.2 Theory

3.2.1 The surface charge and potential

We focus on the equilibrium structure of the interface between a solution and a highly charged and porous electrode [11, 16, 25, 29]. The solution enters the pores and the EDL is formed inside. The total charge Q accumulated per unit mass of active material on the electrode can be obtained from the BET surface area as:

$$Q = \sigma_s \times BET \tag{3.1}$$

where the surface charge density σ_s is obtained either as

$$\sigma_s = -\varepsilon_0 \varepsilon_m \nabla \Psi|_n(\text{surface}) \tag{3.2}$$

or by

$$\sigma_s = \frac{1}{S} \int_{mic} \rho dV \tag{3.3}$$

In Eq. (3.2), $\varepsilon_0 \varepsilon_m$ is the medium permittivity, Ψ the electric potential, and $-\nabla \Psi|_n$(surface) is the normal component of the electric field calculated at the interface electrode/solution. In Eq. (3.3) S is the total surface area in contact with the solution, *mic* refers to the micropore, ρ is the volume charge density, and the integral is made on the whole micropores volume. On the other hand, the charge stored in the EDL is usually characterized by its capacitance, C_d. This can be calculated from the surface charge density by:

$$C_d = \frac{d\sigma_s}{d\Psi_s} \tag{3.4}$$

where Ψ_s is the surface potential.

We simulate the porous electrode as a conductive block with a swarm of cylindrical pores. In this model, an empty Stern layer is included that accounts for the minimum distance of approach to the surface given by the ionic volume. Hence the electric potential profile can be calculated by solving the Poisson equation in the solution inside any of the pores. Accordingly,

$$\nabla^2 \Psi = \begin{cases} -\dfrac{e}{\varepsilon_0 \varepsilon_m}\left(\sum_{i=1}^{N} z^i n^i\right) & 0 < r < R - r^{(1)} \\ -\dfrac{e}{\varepsilon_0 \varepsilon_m}\left(\sum_{i=2}^{(N)} z^i n^i\right) & R - r^{(1)} < r < R - r^{(2)} \\ \cdots & \\ 0 & R - r^{(N)} < r < R \end{cases} \tag{3.5}$$

where e is the electron charge, z^i the valency of ions of species i (sorted by size, starting from the largest), $r^{(i)}$ their radius, n^i their concentration, r the distance to the pore axis, and R is the pore radius. In this way, we define a number of regions in which only ions with the appropriate size can enter. This equation has to be solved with the boundary conditions at the electrode surface, at the limit between adjacent regions and at the pore axis. At the pore surface, the potential must be the surface potential. The second boundary condition emerges from the fact that the EDL overlap in micropores is likely. If this is the case, the electric potential does not reach the zero reference value of the bulk anywhere inside the pore, but the electric field is zero at the pore axis. Hence:

$$\Psi|_{r=R} = \Psi_s \tag{3.6}$$

$$\frac{d\Psi}{dr}|_{r=0} = 0 \tag{3.7}$$

Also, in the boundary between any two regions there must be continuity of the potential.

Finally, we consider that the micropores are coupled to reservoirs of a certain ionic concentration, and hence the equilibrium chemical potential at any point of the solution must be the same as in the reservoirs.

3.2.2 The mD approach

If the pore size is 1 nm or less, then the EDL overlapping leads to a smooth potential decay. A widely used approximation is to consider that these micropores act as charge reservoirs in which a constant potential is established, $\Delta\Psi_D$, larger in absolute value than the electric potential at the unbounded solution [31, 34]. Since in this model (the so-called modified Donnan, mD hereafter) a constant potential is considered, the mathematical problem becomes an easier algebraic set of equations. Accordingly, the (millimolar) concentrations of species i at the micropore $c_{mi,i}$ and at the reservoir $c_{ma,i}$ (also called macropore) are related to this modified Donnan potential through:

$$c_{mi,i} = c_{ma,i}\text{Exp}\left[-z^i \Delta\phi_D - \overline{\mu}^i_{ex}\right] \tag{3.8}$$

where $\Delta\phi_D = e\Delta\Psi_D/k_BT$ and $\overline{\mu}^i_{ex} = \mu^i_{ex}/k_BT$ are the adimensional Donnan potential at the diffuse layer and the adimensional excess chemical potential of species i, respectively; k_B is the Boltzmann constant and T the absolute temperature. Accordingly, the volumetric charge density, ρ, can be calculated with:

$$\rho = F\sum_{i=1}^{N} z^i c_{mi,i} \tag{3.9}$$

where $F = eN_A$ is the Faraday constant and N_A the Avogadro number. In this model, an uncharged Stern layer is considered in series with the diffuse layer, and it is characterized by a volumetric capacitance $C_{v,St}$. The potential drop in this layer, $\Delta\Psi_{St}$, is:

$$\Delta\Psi_{St} = \frac{\rho}{C_{v,St}} \tag{3.10}$$

and the surface potential at the pore wall is:

$$\Psi_s = \Delta\Psi_{St} + \Delta\Psi_D \tag{3.11}$$

3.2.3 The excess chemical potential

The total chemical potential can be divided into three terms accounting for the contribution of (i) the ideal chemical potential for noninteracting particles, μ^i_{id}, (ii) the contribution of the electric interaction, μ^i_{elec}, and (iii) the remaining interactions, μ^i_{ex}:

$$\mu^i_e(\vec{r}) = \mu^i_{id} + \mu^i_{elec} + \mu^i_{ex} \tag{3.12}$$

The first term is:

$$\mu^i_{id} = k_BT\ln[n^i] \tag{3.13}$$

In the frame of the mean field approximation, the electric interactions can be taken into account through the electric energy of every ion in the mean field:

$$\mu^i_{elec} = z^i e\Psi \tag{3.14}$$

These are the only terms in the case of noninteracting ions, that is, in a point-like ion model (PI). If, on the contrary, we consider the contribution of the excluded volume repulsion, we have to add an excess chemical potential. This can be modeled by means of the Bikerman equation [27, 28]:

$$\mu^i_{ex} = k_BT\ln\frac{1}{1-\phi} \tag{3.15}$$

where $\phi = \sum_i \phi_i$ is the volume fraction occupied by all ions and $\phi_i = \frac{4}{3}\pi r^{(i)3} n^i$ is the volume fraction occupied by ions of species i.

This equation can be deduced from a lattice approach and statistical tools. Another possibility is to consider the equation of state of a gas of rigid spheres. In the C-S model [26, 30, 35] this equation can be accurately simplified and, from this simplification, an excess chemical potential can be calculated:

$$\mu_{ex}^{i} = k_B T \frac{\phi(8 - 9\phi + 3\phi^2)}{(1-\phi)^3} \tag{3.16}$$

This model predicts EDL crowding only at very large surface potentials, while the Bikerman equation predicts the saturation of the EDL at moderate potentials. This means that with the C-S model, the steric repulsion is not a short-distance interaction as in the case of the Bikerman equation.

When dealing with multiionic solutions, the equation of state of BMCSL [17, 32, 33] has been shown to be more appropriate. This equation takes into account that ions with equal charge but different sizes behave differently. The excess chemical potential for species i is given by:

$$\begin{aligned} \frac{\mu_{ex,i}}{k_B T} &= -\left(1 + \frac{2\zeta_2^3 d_i^3}{\phi^3} - \frac{3\zeta_2^2 d_i^2}{\phi^2}\right) \ln(1-\phi) \\ &+ \frac{3\zeta_2 d_i + 3\zeta_1 d_i^2 + \zeta_0 d_i^3}{1-\phi} + \frac{3\zeta_2 d_i^2}{(1-\phi)^2}\left(\frac{\zeta_2}{\phi} + \zeta_1 d_i\right) \\ &- \zeta_2^3 d_i^3 \frac{\phi^2 - 5\phi + 2}{\phi^2 (1-\phi)^3} \end{aligned} \tag{3.17}$$

where $d_i = 2r^{(i)}$ is the ionic diameter and $\zeta_k = \sum_i \phi_i d_i^{k-3}$.

A more detailed analysis of the possible approaches to the excess chemical potential can be found in [30, 35].

3.3 EDL Structure: Bikerman Equation vs C-S Model

In Fig. 3.1A we compare the predictions of the electric potential profile inside a 2-nm pore when PI, Bikerman (Bik), or C-S models are used. In general, the PI model predicts a deep decay and nonoverlapping EDLs, while in the case of the Bikerman equation the potential decay is slower. Finally, unlike C-S model, Bik model underestimates excluded volume effects. Hence, a slower decay of the electric potential is observed by using the former model. In fact, at moderately low surface potentials, EDL overlap is observed (this manifests in a nonzero potential at the pore axis). As a consequence, the counterion concentration (Fig. 3.1B), the surface charge density (Fig. 3.1C), and the EDL capacitance (Fig. 3.1D) are smaller. Given the different predictions that are expected with the three models, hereafter we will use the C-S model, unless otherwise indicated.

This result is in contrast to that in [31], where the author obtains small differences with respect to the PI approximation. In their approach, the author

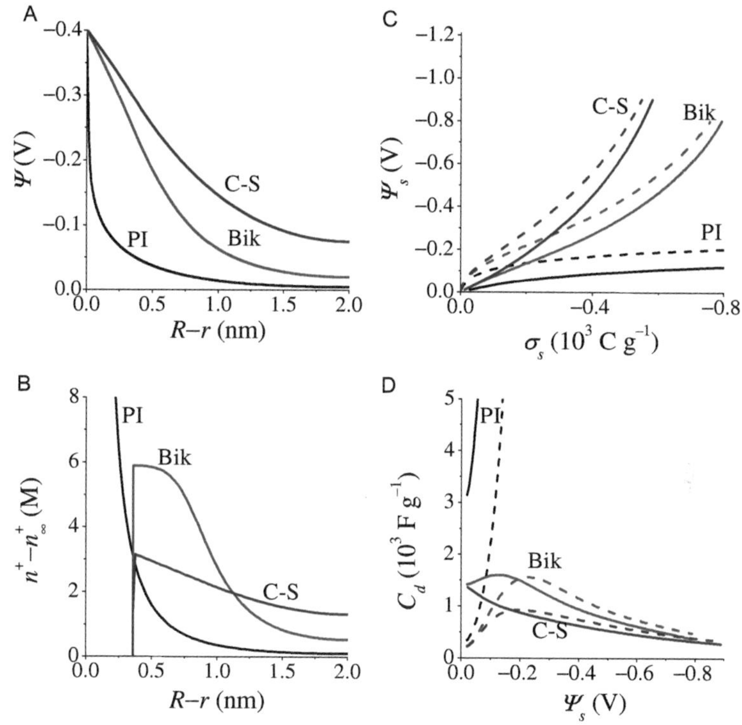

FIG. 3.1 Electric potential (A) and excess concentration of Na^+ ions (B) as a function of the distance to the pore surface for $\Psi_s = -0.4$ V. NaCl in the bulk: 500 mM. (C) Relation between the surface potential and surface charge per unit mass of electrode in presence of 20 mM (*dashed lines*) and 500 mM (*solid lines*) NaCl solution. (D) Differential capacitance per unit mass of electrode as a function of the surface potential in presence of 20 mM (*dashed lines*) and 500 mM (*solid lines*) NaCl solution. In all graphs, PI, Bikerman, and C-S models are used. Pore radius: 2 nm. BET area 2000 m^2/g. *(Reproduced with permission of the PCCP Owner Societies, Jiménez ML, Ahualli S, Arenas-Guerrero P, Fernández MM, Iglesias G, Delgado AV. Multiionic effects on the capacitance of porous electrodes. Phys Chem Chem Phys 2018;20:5012–20.)*

uses an mD model, in which the experimental value 0.2 GF/m^3 is used for the volumetric capacitance of the Stern layer. This corresponds, in the case of a 2-nm cylindrical pore radius, to a capacitance of $C_{St} = 0.2$ F/m^2. In our case, the Stern layer is a water layer of the thickness of the ionic radius, and hence the capacitance is $C_{St} = 2$ F/m^2, 10 times larger. As a consequence, the potential decay in the diffuse layer will be larger in our approach, and hence ionic accumulation and steric repulsion will be more significant. Furthermore, although the mD approximation is often used when describing the electric potential distribution inside micropores, it must be mentioned that, with this approximation, one neglects that close to the electrode surface, the ionic concentration in the EDL is larger than the average value, leading to further underestimation of the steric repulsion and subsequently to overestimation of the total charge in the EDL [29].

The steric repulsion prevents large accumulations of ions in the EDL, and as a consequence the surface charge saturates (see Fig. 3.2A). This is manifested in a maximum value and subsequent decrease of the EDL capacitance C_d (Fig. 3.2B). This maximum takes place at lower surface potential the larger the ionic strength. As a consequence, above certain surface potential there is a characteristic reversal of the capacitance-ionic strength relationship: the capacitance at 500 mM becomes smaller than at 20 mM, that is, $\Delta C_d = C_d$ (500 mM) − C_d (20 mM) becomes negative. This means that the curves corresponding to 500 and 20 mM in Fig. 3.2A approach each other. This has an important effect in CDLE, since it is sustained by the difference between both curves, and hence saturation limits the energy that can be obtained with this technique [16, 25, 29].

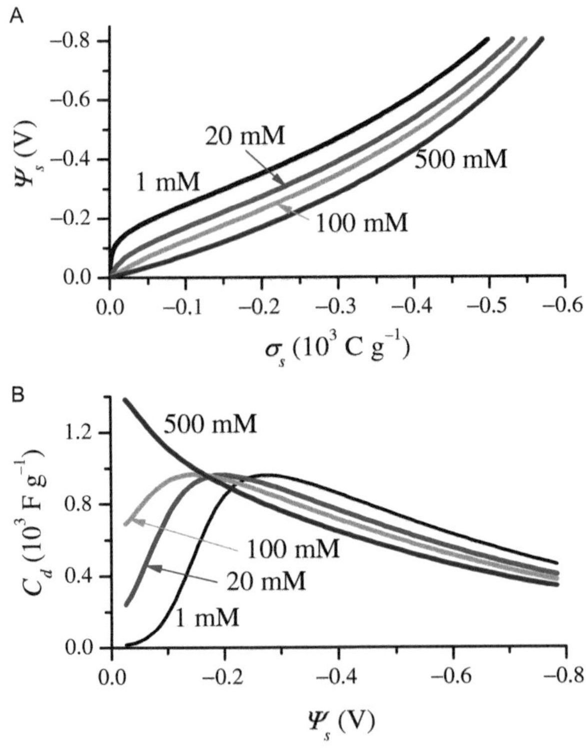

FIG. 3.2 (A) Surface potential as a function of the surface charge density per unit mass of electrode for the NaCl ionic concentration indicated. (B) Differential capacitance per unit mass of electrode as a function of the surface potential for the same cases. Pore radius: 2 nm. BET area 2000 m^2/g.

3.4 Ion Specificity

3.4.1 Effect of valency

In Fig. 3.3A and B we show the excess ionic concentration and potential profiles for low surface potential. We can see that the curves corresponding to 20 and 500 mM ionic strength exhibit large differences between each other and hence large differences are expected in their capacitances. At large surface potentials (Fig. 3.3C and D) steric repulsion prevents large ionic accumulations. We can see that the profiles for different ionic strengths become very similar. This effect is more relevant for larger valencies. In fact, we can see that at $z = 3$ the profiles are nearly independent of the ionic strength.

Hence it can be said that the larger the valency, the lower the potential at which saturation is reached. This is reflected in Fig. 3.4B where we can see

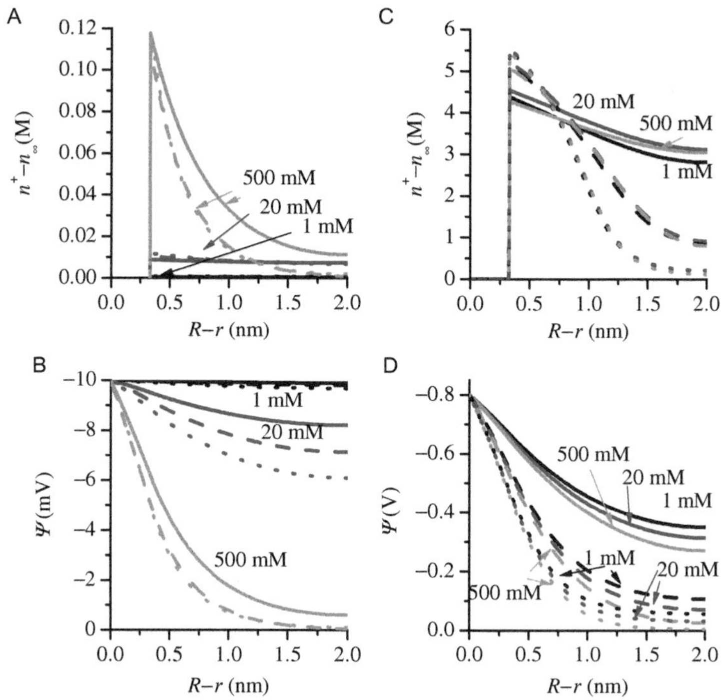

FIG. 3.3 (A) Excess of ionic concentration as a function of the distance to the surface of an electrode immersed in a solution 1 mM, 20 mM, and 500 mM of ACl_z, with *A* being the cation and its valency $z = 1$ (*solid lines*), 1 (*dashed lines*), and 3 (*dotted lines*). Surface potential: −0.01 V; (B) Electric potential as a function of the distance to the surface for the same solutions as in (A). (C) and (D) Same as (A) and (B) but for surface potential −0.8 V. Pore radius: 2 nm. Cation size: 0.358 nm.

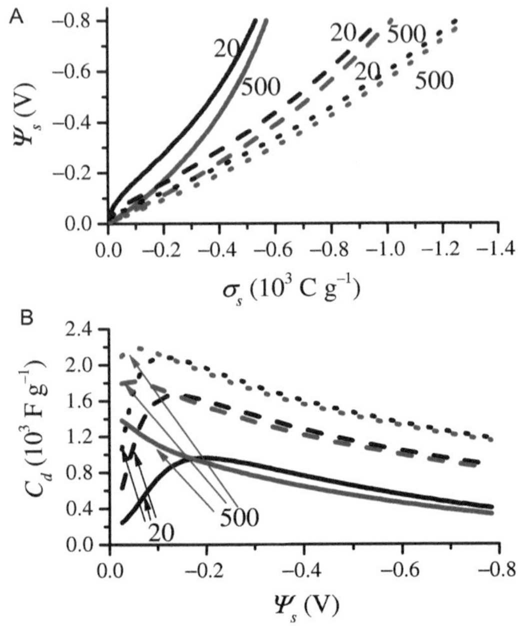

FIG. 3.4 (A) Surface potential as a function of the surface charge density per unit mass of electrode in presence of 20 mM and 500 mM of ACl_z solutions, with *A* being the cation and its valency $z = 1$ (*solid lines*), 1 (*dashed lines*), and 3 (*dotted lines*). (B) Differential capacitance per unit mass of electrode as a function of the surface potential for the same cases. Pore radius: 2 nm. BET area 2000 m^2/g, Cation size: 0.358 nm.

that the surface potential of the reversal of the capacitance-ionic strength relationship is lower for larger valencies. As a consequence, also the EDL expansion upon salinity exchange is smaller in this case (see Fig. 3.4A).

3.4.2 Effect of the ionic size

Fig. 3.5 illustrates the effect of ionic size on the EDL structure. As expected, since steric repulsion is only important at sufficiently large ionic concentrations, differences are visible for large ionic strengths and/or potentials (Fig. 3.5C and D). In these cases, the potential decay in the EDL is smoother for larger ions, and as a consequence, the surface charge density and the differential capacitance are smaller (see Fig. 3.6).

3.5 EDL Overlap

Another important aspect to take into account in the EDL structure is the role of the EDL overlap. Micropores are very narrow and hence there is not enough space to screen the surface potential. This is observed as a nonzero potential at the pore axis. Note that the Debye length

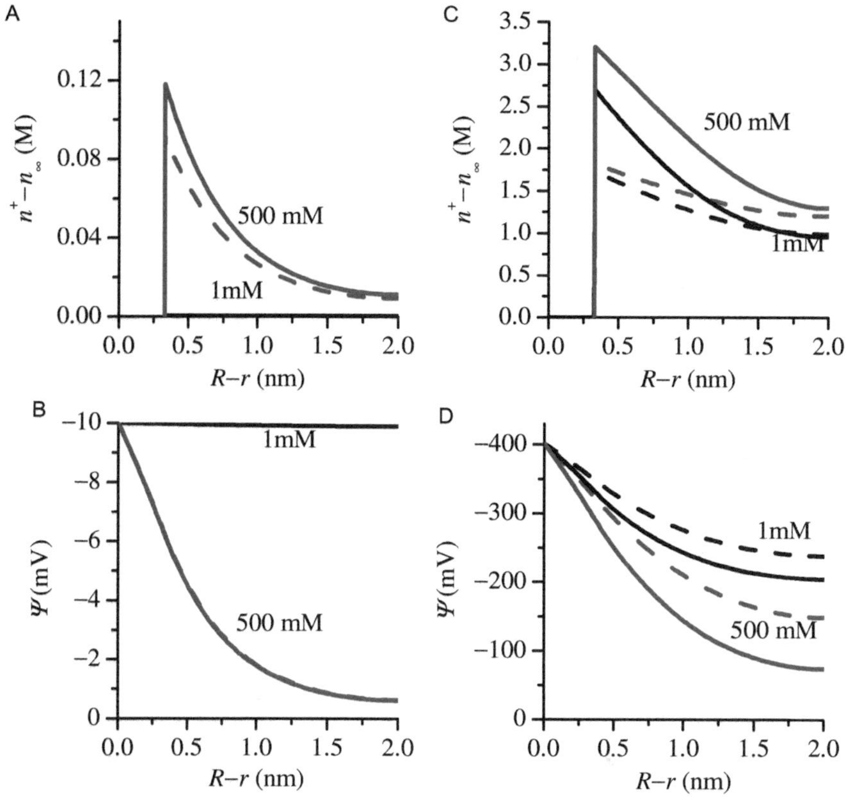

FIG. 3.5 (A) Excess of ionic concentration as a function of the distance to the surface of an electrode immersed in a solution 1 mM and 500 mM of ACl, with *A* being a monovalent cation of size $r^+ = 0.358$ nm (*solid lines*) and 0.428 nm (*dashed lines*). Surface potential −0.01 V; (B) electric potential as a function of the distance to the surface for the same solutions as in (A). (C) and (D) Same as (A) and (B) but for surface potential −0.4 V. Pore radius: 2 nm.

$$\kappa^{-1} = \sqrt{\frac{\varepsilon_0 \varepsilon_m k_B T}{\sum_{j=1}^{N} \left(z^j\right)^2 e^2 n_\infty^j}} \tag{3.18}$$

is roughly 2 and 0.4 nm for 20 and 500 mM, respectively, and independent of the surface potential. These values do not justify the slow potential decay in Figs. 3.3 and 3.5. With steric repulsion, the EDL length, as the length at which the potential roughly decays to Ψ_s/e, is no longer represented by the Debye length in Eq. (3.18). Moreover, if we compare Fig. 3.3B and D, we can see that EDL thickness becomes dependent of the surface potential.

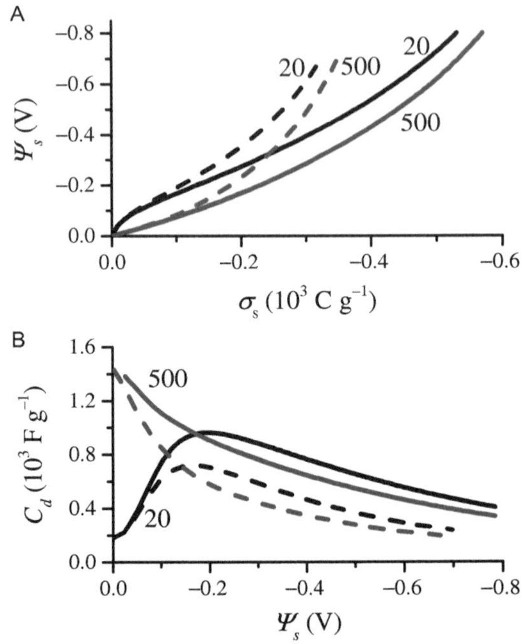

FIG. 3.6 (A) Surface potential versus surface charge density per unit mass of electrode in presence of 20 mM and 500 mM of ACl, with *A* being a monovalent cation of size $r^+ = 0.358$ nm (*solid lines*) and 0.428 nm (*dashed lines*). (B) Differential capacitance per unit mass of electrode as a function of the surface potential for the same cases. Pore radius: 2 nm. BET area 2000 m^2/g.

In the previously mentioned figures we observe that there are situations in which the potential decay is less than half its value. A possible approach in these cases is to consider an mD model [19, 31, 34, 36]. However, as observed in these figures, even for small pores we can find situations in which such a description of the EDL profiles is not accurate. In fact, at low surface potential and high ionic strengths (Fig. 3.5B) or at large surface potentials (Fig. 3.5D), the EDL overlap is absent even for micropores. Hence solving the Poisson equation to obtain the potential profile may give a more accurate picture of the EDL behavior.

We can get deeper information on the EDL overlap by exploring the different pore sizes. In Fig. 3.7 we can see that the larger the pore size, the larger the surface charge density and the capacitance. This is due to the fact that at the pore axis no electric field is expected, and this limits the extent to which the electric field, hence the charge, can reach at the pore surface. It is worthy to note that in this figure the BET area is supposed to be the same. However, large BET areas are found for smaller pore sizes. Hence a good balance for having large capacitances must take both aspects, BET area and EDL overlap, into account.

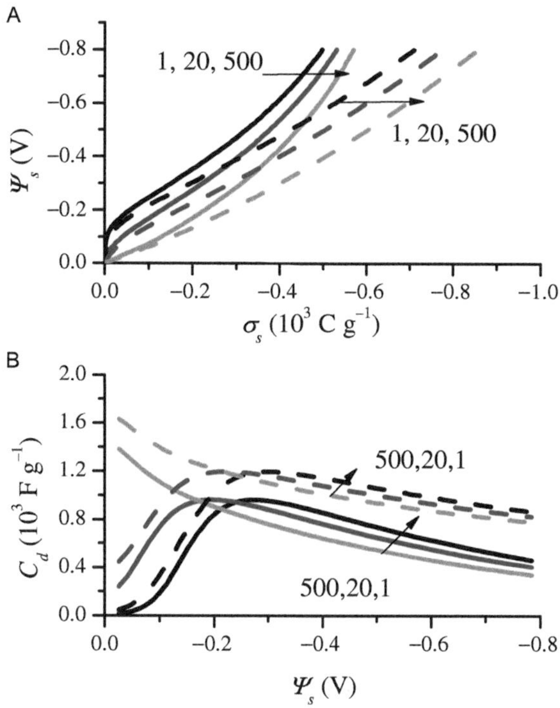

FIG. 3.7 (A) Surface potential versus surface charge density per unit mass of electrode for pore radius 2 nm (*solid lines*) and 5 nm (*dashed lines*) and the NaCl concentration indicated. (B) Differential capacitance per unit mass of electrode as a function of the surface potential for the same cases. BET area 2000 m^2/g.

3.6 Multiionic Solutions

3.6.1 Competition between ions with different valency

When there are two or more ions with the same valency in the solution, the BMCSL equation is more appropriate than the C-S one, since it captures the fact that the free energy stored in the EDL is smaller when the concentration of the smaller ions is larger than the big ones. Even in the case of a mixture of (small) monovalent ions and (large) divalent ions, it has been shown [17] that the concentration of the former close to the surface can be higher than that of the latter. This means that, in this case, steric effects predominate over electrostatic ones. However, a sufficiently large surface charge is needed for this to occur, as it was, in fact, experimentally observed in [37] and theoretically justified in [38]. If we restrict ourselves to surface potentials below 0.8 V, the surface charge density will be below 0.3 C/m^2. In this regime, the steric effects are not pronounced enough to overcome the valency effect [17]. This is illustrated in Fig. 3.8, where the ionic concentration is plotted as a function

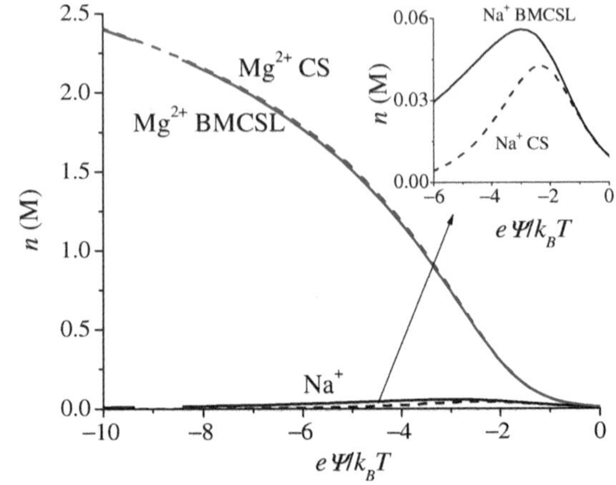

FIG. 3.8 Na^+ and Mg^{2+} concentrations as a function of the electric potential for an interface in contact with a reservoir 10 mM Na^+ and 10 mM Mg^{2+}. *Solid lines*: BMCSL equation of state. *Dashed lines*: C-S equation of state. *Inset*: detail of the Na^+ concentration. *(Reproduced with permission of the PCCP Owner Societies, Jiménez ML, Ahualli S, Arenas-Guerrero P, Fernández MM, Iglesias G, Delgado AV. Multiionic effects on the capacitance of porous electrodes. Phys Chem Chem Phys 2018;20:5012–20.)*

of the electric potential for a solution in contact with a reservoir 10 mM Na^+ and 10 mM Mg^{2+}. It can be seen that, in the potential range studied, both C-S and BMCSL equations of state predict nearly the same Mg^{2+} concentration, whereas that of Na^+ is underestimated by the C-S equation. However, due to steric repulsion, this concentration is negligible, and hence the capacitance will be dominated by divalent ions.

Accordingly, in multiionic (MI hereafter) solutions, ions with the larger valency occupy the EDL. If the ionic size is not taken into account, the other ions still can reside near the surface. However, for large potentials and/or ionic strengths, the EDL crowding prevents the accumulation of the weakest ions. This is illustrated in Fig. 3.9, where the EDL composition is plotted for a mixture of NaCl 0.0163 M and $MgCl_2$ 0.0019 M solutions. This value was chosen to preserve the ratio between the concentrations of monovalent and divalent cations similar to the typical sea values [25]. At low potentials (Fig. 3.9A) the EDL in the mixture is mainly composed of Na^+ ions (dashed line) as expected from the higher Na^+ content in the bulk. As the potential increases, the volume available is occupied by Mg^{2+}, giving a minor role to Na^+ ions (Fig. 3.9B).

The presence of divalent ions leads to a deeper decay of the electric potential (see Fig. 3.9C), and hence a larger surface charge density than that expected from the large Na^+ content (Fig. 3.10A). However, as already discussed, when the EDL is formed by divalent ions, ΔC_d is smaller than in the case of monovalent ions (see Fig. 3.10B), similar to the Bikerman model's predictions in [25].

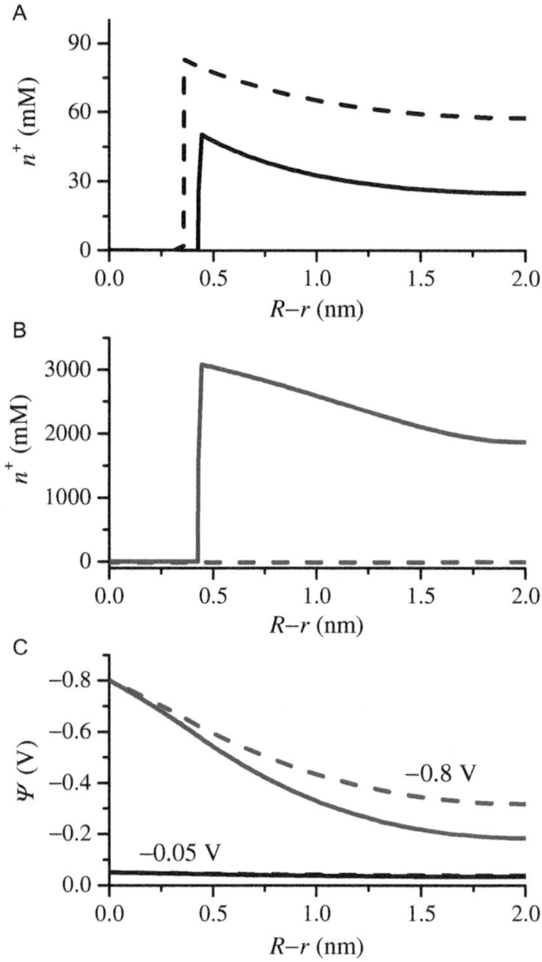

FIG. 3.9 (A) Na^+ (*dashed lines*) and Mg^{2+} (*solid lines*) concentrations as a function of the distance to the surface of an electrode immersed in a 20 mM MI solution and $\Psi_s = -0.05$ V. (B) Same as (A) but for $\Psi_s = -0.8$ V. (C) Electric potential profile in the same conditions for 20 mM NaCl (*dashed line*) and 20 mM MI (*solid line*) solutions. Surface potentials: −0.05 V and −0.8 V. Pore radius 2 nm. *(Reproduced with permission of the PCCP Owner Societies, Jiménez ML, Ahualli S, Arenas-Guerrero P, Fernández MM, Iglesias G, Delgado AV. Multiionic effects on the capacitance of porous electrodes. Phys Chem Chem Phys 2018;20:5012–20.)*

3.6.2 Competition between ions with equal valency

Steric repulsion determines not only the EDL composition for unequal valencies multiionic solutions. CDI experiments suggest that there is a certain degree of selectivity between ions with the same valency in the EDL adsorption [22, 23, 39–41]. This fact can be understood in terms of the free energy stored

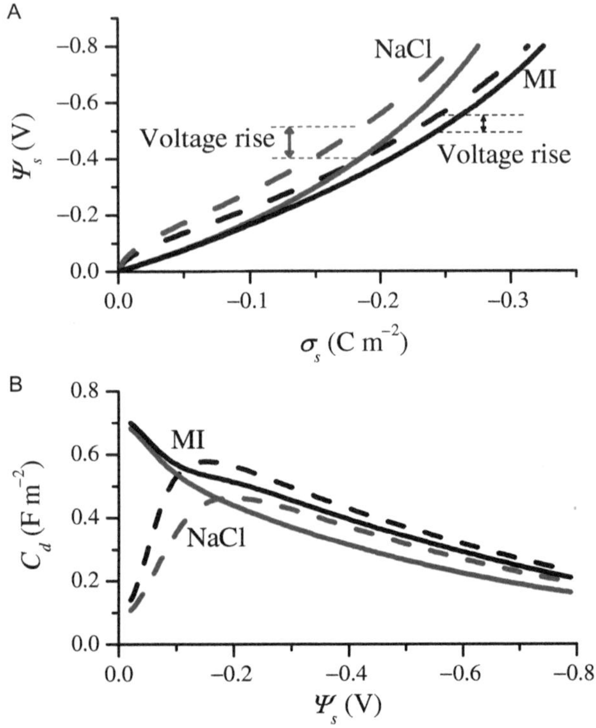

FIG. 3.10 (A) Surface potential-surface charge density relation, and (B) differential capacitance as a function of the surface potential, for NaCl solutions and MI solutions. *Dashed lines*: 20 mM; *solid lines*: 500 mM. Pore radius 2 nm. The *arrows* in (A) indicate the voltage rise upon salinity exchange. *(Reproduced with permission of the PCCP Owner Societies, Jiménez ML, Ahualli S, Arenas-Guerrero P, Fernández MM, Iglesias G, Delgado AV. Multiionic effects on the capacitance of porous electrodes. Phys Chem Chem Phys 2018;20:5012–20.)*

in the EDL. Electrosorption of the smallest species occurs because this provides the maximum configurational entropy, and hence minimum free energy.

We can use an MI solution 10 mM KCl and 10 mM NaCl inside a 2-nm micropore as an example. In this case, strong EDL overlap is expected. Thus, the mD model in combination with a BMCSL equation of states can give us a simple approach [31]. In Fig. 3.11A it is shown that for low surface potentials no difference between Na^+ and K^+ concentrations is expected. However, for potentials above 0.2 V, the small difference in size between Na^+ and K^+ ions is enough to produce an odd concentration of these ions inside the micropore. As a consequence, some deviations in the surface charge density are observed (Fig. 3.11B). At low potentials, micropores in the presence of MI solution store larger charge density as both Na^+ and K^+ are present. On the other hand, at large potentials, K^+ ions determine the micropore charge (Fig. 3.11B) and capacitance (Fig. 3.11C).

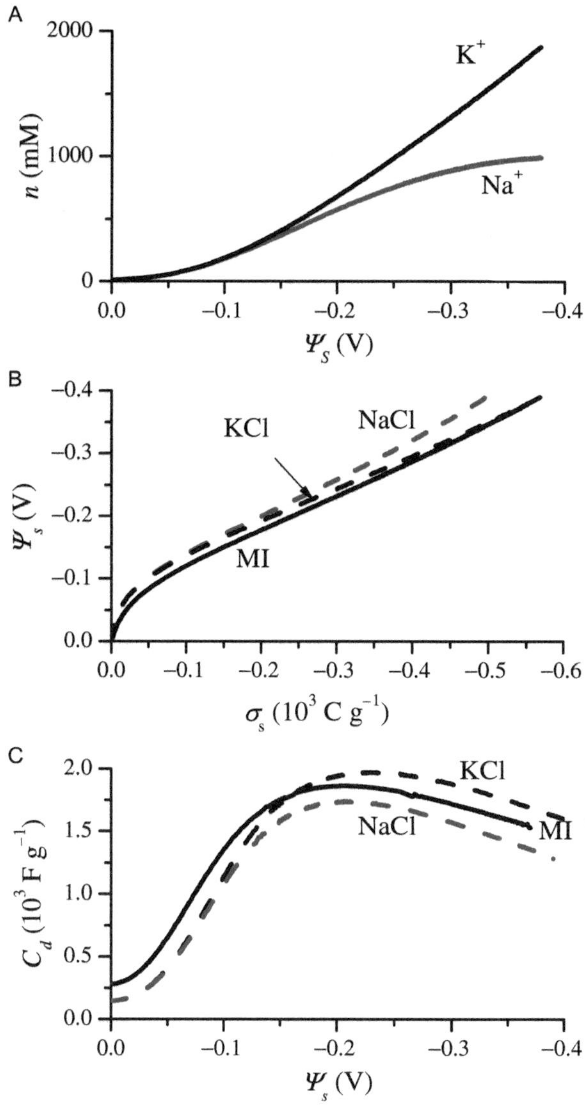

FIG. 3.11 (A) Counterion concentration inside a micropore immersed in an MI solution 10 mM KCl and 10 mM NaCl as a function of the surface potential. (B) Surface potential as a function of the surface charge density, and (C) differential capacitance as a function of the surface potential, for the same solution (*solid line*), for a 10 mM KCl solution (*dashed line*), and for 10 mM NaCl (*dashed line*). Pore radius 2 nm. BET area 2000 m^2/g. Stern layer volumetric capacitance $C_{v,St} = 2$ GF/m^3.

3.6.3 Size effects: beyond steric repulsion

The ionic size has been until now taken into account as a source of steric repulsion. However, if ions are envisaged as finite size spheres, they are not expected to have the same dielectric properties as the water. For instance, in [42] the authors treat ions as dielectric spheres nonhomogeneously distributed. As a consequence, in addition to the steric repulsion, two other forces related to the finite size of the ions appear:

- The Born force: The inhomogeneous concentration produces a variation of the electric permittivity across the solution. The Born force appears because ions tend to move to regions of higher permittivity, that is, away from the EDL.
- The dielectrophoretic force: Ions are dielectric spheres, and hence the electric field in the EDL polarizes them.

By solving the Poisson equation, the authors calculate the capacitance of the EDL beside a charged flat surface immersed in an unbounded electrolyte. An example is shown in Fig. 3.12, where they compare the theoretical predictions of the Bikerman equation with the C-S one with and without the dielectrophoretic and Born forces and with the experimental results in [43], that could not be previously explained. In this figure it can be seen the improvement that can be achieved by using C-S equation of states instead of Bikerman equation. But the most relevant result is the determinant role of Born and dielectrophoretic forces for surface potentials as low as 100 mV.

3.7 Steric Effect in CDLE

CDLE [3] is based on the following idea: the capacitance of the EDL close to the charged electrode increases with the ionic strength, and hence the charge

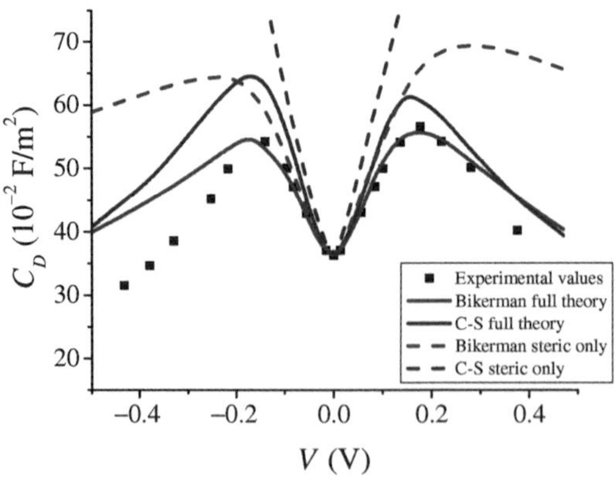

FIG. 3.12 Dependence of the total differential capacitance (inner and diffuse parts of the electric double layer) on the potential drop across this layer. Comparison with experimental data [43] of the theoretical predictions calculated for KPF_6 aqueous electrolyte solutions.

of the electrode in the presence of salty water increases its potential energy when salty and fresh waters are exchanged (the capacitance decreases). The total work that can be extracted in one cycle is given by the area enclosed between two curves in Fig. 3.2A (usually fresh 20 mM and salty 500 mM solutions). This area mainly depends on the charge stored in the micropores, that is, their capacitance, and the voltage rise, that is, the vertical distances between these two curves. Hence a good strategy is to increase the working voltage of the electrodes, since by doing so, the surface charge is increased. However, the voltage rise is determined by the dependence of the capacitance on the ionic strength. That is, the larger $\Delta C_d = C_d(\text{salty}) - C_d(\text{fresh})$ the larger energy that can be obtained. As we can see in Fig. 3.2B, ΔC_d becomes negative, and for larger surface potentials smaller energies can be obtained. As a consequence, the extracted work W exhibits a maximum. It is worthy to point out that the sign reversal in ΔC_d occurs due to steric effects. It is also relevant to note that it always occurs, whatever the pair of ionic strength that are chosen (Fig. 3.2B), the only difference being the potential at which it takes place.

Fig. 3.13 presents the extracted work per unit surface area of the positive and negative electrodes (Fig. 3.13A and B, respectively) and the total extracted work for the cases of 2 and 20 nm pores. The larger extracted work in the latter case is due to the reduced EDL overlap and subsequent larger surface charge density. However, it must be taken into account that generally, the larger the pore, the smaller the BET area is, and hence the extracted work per unit mass of electrode, as observed in Fig. 3.13C. In this figure, the x-axis is the voltage difference between both electrodes. Note that since the electrolyte is not generally symmetrical, neither are the EDLs at both electrodes. This

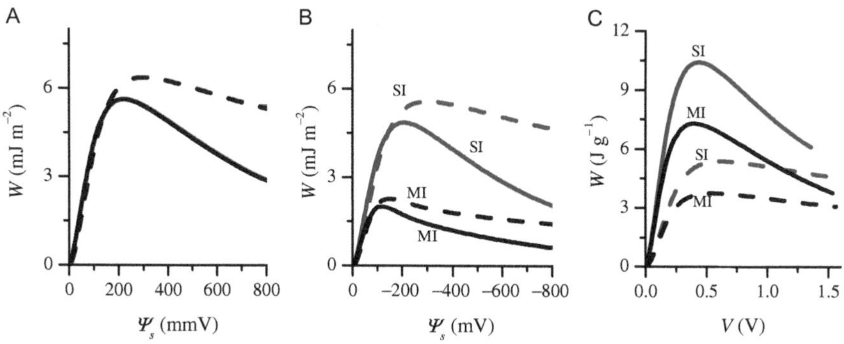

FIG. 3.13 Extracted energy per cycle and per unit surface area in the positive (A) and negative (B) electrodes as a function of the surface potential. (C) Extracted energy per cycle and unit electrode mass as a function of the cell potential. Pore sizes: 2 nm (*solid lines*) and 20 nm (*dashed lines*). Salts used: SI (20 and 500 mM NaCl) and MI. In MI solutions the anion is 20 and 500 mM Cl^-, and the cations are Mg^{2+} and Na^+ at a ratio 1/10. *(Reproduced with permission of the PCCP Owner Societies, Jiménez ML, Ahualli S, Arenas-Guerrero P, Fernández MM, Iglesias G, Delgado AV. Multiionic effects on the capacitance of porous electrodes. Phys Chem Chem Phys 2018;20:5012–20.)*

situation is especially important in the case of multiionic solutions, and hence the voltage difference must be calculated as:

$$V(\sigma) = \Psi_s^+(\sigma) - \Psi_s^-(-\sigma) \tag{3.19}$$

with $\Psi_s^{+/-}$ being the potential at the positive/negative electrode and σ the surface charge density.

If we compare SI and MI solutions, the relevant effects take place in the negative electrode, where EDL is composed of cations. It can be seen that the presence of divalent ions decreases the amount of energy that can be extracted. In Fig. 3.3 it was observed that the EDL is mainly composed of divalent ions, which, in turn, produce an increase of the stored charge (Fig. 3.4). However, ΔC_d decreases, and hence the voltage rises. At the same time, Mg^{2+} ions are larger than Na^+ ones, and hence the stored charge does not increase so much as expected from the double valency. The result is that the extracted energy is greatly reduced.

However, the strong variation of the capacitance with the presence of the divalent Mg^{2+} ions can be taken advantage of. Accordingly, a CDLE cycle may be designed in which the salty water solution is a 500 mM MI solution, while the fresh one is a 20 mM NaCl solution. This is schematically shown in Fig. 3.14. In this figure we can see that for a given surface charge density, the potential rise should be larger in this case. Accordingly, we can conclude that the CDLE process could be more profitable if a location can be found where salty waters containing divalent ions are used jointly with an SI fresh solution. If necessary, resins can be used to get rid of multivalent ions from fresh water.

A different strategy was analyzed in [11]. In that case the temperature of the fresh water was raised, thus decreasing its capacitance. The reverse of

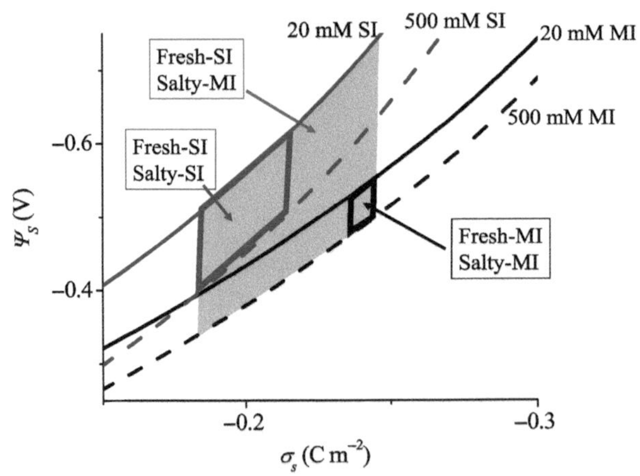

FIG. 3.14 Same as Fig. 3.10A but with a highlighted CDLE cycle. In the graph, the cycle with SI solutions, MI solutions and mixed cycle are highlighted. *(Reproduced with permission of the PCCP Owner Societies, Jiménez ML, Ahualli S, Arenas-Guerrero P, Fernández MM, Iglesias G, Delgado AV. Multiionic effects on the capacitance of porous electrodes. Phys Chem Chem Phys 2018;20:5012–20.)*

ΔC_d can be avoided (see Fig. 3.15A) and a monotonic increase of the energy with the working potential can be achieved (see Fig. 3.15B). This possibility was also experimentally successfully explored (Fig. 3.15C), despite leakage and an unsuccessful micropores wetting preventing a quantitative matching.

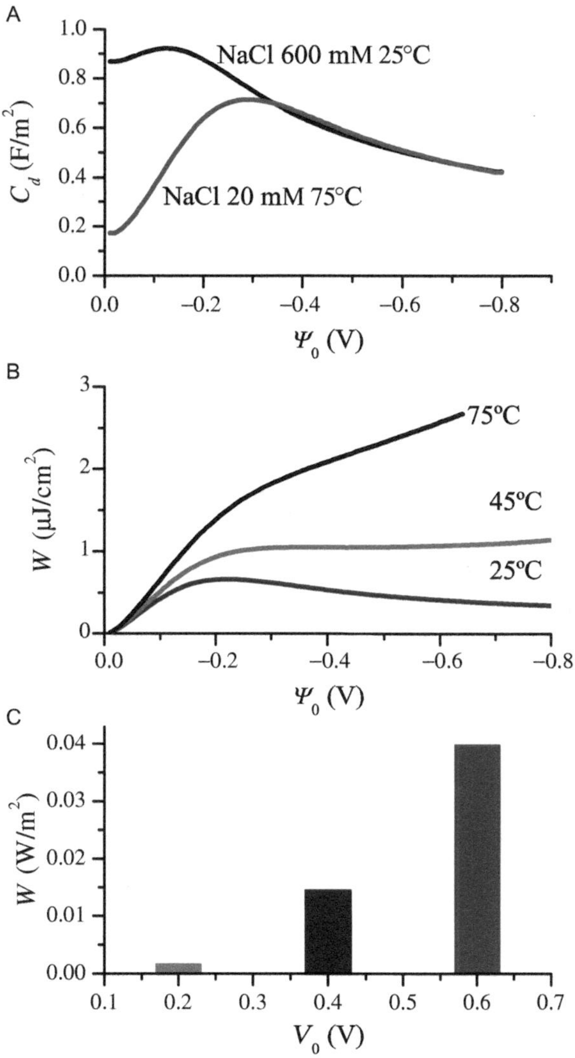

FIG. 3.15 (A) Theoretical predictions for the capacitance of an electrode with 5-nm micropores immersed in 20 mM (75°C) and 600 mM (25°C) NaCl solutions. (B) Theoretical predictions for the extracted energy in a CDLE cycle per unit surface area, with the same electrode in which the fresh solution is heated at the temperature indicated. (C) Experimental extracted energy per unit electrode surface. More details can be found in [11]. *(Reprinted with permission from Ahualli S, Fernández MM, Iglesias G, Delgado AV, Jiménez ML. Temperature effects on energy production by salinity exchange. Environ Sci Technol 2014:48:12378–85. Copyright 2014 American Chemical Society.)*

Acknowledgments

Financial support of this investigation by University of Granada (grant Plan Propio de Investigación) and Junta de Andalucía, Spain (grant No. PE2012-FQM0694) is gratefully acknowledged.

References

[1] Pandolfo AG, Hollenkamp AF. Carbon properties and their role in supercapacitors. J Power Sources 2006;157(1):11–27.

[2] Merlet C, Rotenberg B, Madden PA, Taberna PL, Simon P, Gogotsi Y, et al. On the molecular origin of supercapacitance in nanoporous carbon electrodes. Nat Mater 2012;11(4):306–10.

[3] Brogioli D. Extracting renewable energy from a salinity difference using a capacitor. Phys Rev Lett 2009;103(5):058501.

[4] Boon N, van Roij R. "Blue energy" from ion adsorption and electrode charging in sea and river water. Mol Phys 2011;109(7–10):1229–41.

[5] Bijmans MFM, Burheim OS, Bryjak M, Delgado AV, Hack P, Mantegazza F, et al. Capmix-deploying capacitors for salt gradient power extraction. Energy Procedia 2012;20:108–15.

[6] Marino M, Misuri L, Jiménez ML, Ahualli S, Kozynchenko O, Tennison S, et al. Modification of the surface of activated carbon electrodes for capacitive mixing energy extraction from salinity differences. J Colloid Interface Sci 2014;436:146–53.

[7] Anderson MA, Cudero AL, Palma J. Capacitive deionization as an electrochemical means of saving energy and delivering clean water. Comparison to present desalination practices: will it compete? Electrochim Acta 2010;55(12):3845–56.

[8] Porada S, Zhao R, van der Wal A, Presser V, Biesheuvel PM. Review on the science and technology of water desalination by capacitive deionization. Prog Mater Sci 2013;58(8):1388–442.

[9] Suss ME, Porada S, Sun X, Biesheuvel PM, Yoon J, Presser V. Water desalination via capacitive deionization: what is it and what can we expect from it? Energy Environ Sci 2015;8:2296–319.

[10] Wang L, Biesheuvel PM, Lin S. Reversible thermodynamic cycle analysis for capacitive deionization with modified Donnan model. J Colloid Interface Sci 2018;512(Suppl C):522–8. https://doi.org/10.1016/j.jcis.2017.10.060. http://www.sciencedirect.com/science/article/pii/S0021979717312262.

[11] Ahualli S, Fernández MM, Iglesias G, Delgado AV, Jiménez ML. Temperature effects on energy production by salinity exchange. Environ Sci Technol 2014;48(20):12378–85.

[12] Iglesias GR, Fernández MM, Ahualli S, Jiménez ML, Kozynchenko OP, Delgado AV. Materials selection for optimum energy production by double layer expansion methods. J Power Sources 2014;261:371–7.

[13] Chao L, Liu Z, Zhang G, Song X, Lei X, Noyong M, et al. Enhancement of capacitive deionization capacity of hierarchical porous carbon. J Mater Chem A 2015;3(24):12730–7.

[14] Chen JH, Li WZ, Wang DZ, Yang SX, Wen JG, Ren ZF. Electrochemical characterization of carbon nanotubes as electrode in electrochemical double-layer capacitors. Carbon 2002;40(8):1193–7.

[15] Xu Y, Lin Z, Zhong X, Huang X, Weiss NO, Huang Y, et al. Holey graphene frameworks for highly efficient capacitive energy storage. Nat Commun 2014;5:4554.

[16] Jiménez ML, Fernández MM, Ahualli S, Iglesias GR, Delgado AV. Predictions of the maximum energy extracted from salinity exchange inside porous electrodes. J Colloid Interface Sci 2013;402:340–9.
[17] Biesheuvel PM, van Soestbergen M. Counterion volume effects in mixed electrical double layers. J Colloid Interface Sci 2007;316(2):490–9.
[18] Biesheuvel PM. Thermodynamic cycle analysis for capacitive deionization. J Colloid Interface Sci 2009;332(1):258–64. https://doi.org/10.1016/j.jcis.2008.12.018.
[19] Biesheuvel PM, Porada S, Levi M, Bazant MZ. Attractive forces in microporous carbon electrodes for capacitive deionization. J Solid St Electrochem 2014;18(5):1365–76. https://doi.org/10.1007/s10008-014-2383-5.
[20] Giera B, Henson N, Kober EM, Shell MS, Squires TM. Electric double-layer structure in primitive model electrolytes: comparing molecular dynamics with local-density approximations. Langmuir 2015;31(11):3553–62. https://doi.org/10.1021/la5048936.
[21] Seo SJ, Jeon H, Lee JK, Kim GY, Park D, Nojima H, et al. Investigation on removal of hardness ions by capacitive deionization (CDI) for water softening applications. Water Res 2010;44(7):2267–75.
[22] Hou CH, Huang CY. A comparative study of electrosorption selectivity of ions by activated carbon electrodes in capacitive deionization. Desalination 2013;314:124–9.
[23] Li Y, Zhang C, Jiang Y, Wang TJ, Wang H. Effects of the hydration ratio on the electrosorption selectivity of ions during capacitive deionization. Desalination 2016;399:171–7.
[24] Zhao R, van Soestbergen M, Rijnaarts HHM, van der Wal A, Bazant MZ, Biesheuvel PM. Time-dependent ion selectivity in capacitive charging of porous electrodes. J Colloid Interface Sci 2012;384:38–44. https://doi.org/10.1016/j.jcis.2012.06.022.
[25] Fernández MM, Ahualli S, Iglesias GR, González-Caballero F, Delgado AV, Jiménez ML. Multi-ionic effects on energy production based on double layer expansion by salinity exchange. J Colloid Interface Sci 2015;446:335–44.
[26] Carnahan NF, Starling KE. Equation of state for nonattracting rigid spheres. J Chem Phys 1969;51(2):635–6.
[27] Adamczyk Z, Warszynski P. Role of electrostatic interactions in particle adsorption. Adv Colloid Interface Sci 1996;63:41–3149.
[28] Borukhov I. Charge renormalization of cylinders and spheres: ion size effects. J Polymer Sci Part B-Polymer Phys 2004;42(19):3598–615.
[29] Jiménez ML, Ahualli S, Arenas-Guerrero P, Fernández MM, Iglesias G, Delgado AV. Multiionic effects on the capacitance of porous electrodes. Phys Chem Chem Phys 2018;20:5012–20.
[30] López-García JJ, Aranda-Rascón MJ, Grosse C, Horno J. Equilibrium electric double layer of charged spherical colloidal particles: effect of different distances of minimum ion approach to the particle surface. J Phys Chem B 2010;114(22):7548–56.
[31] Suss ME. Size-based ion selectivity of micropore electric double layers in capacitive deionization electrodes. J Electrochem Soc 2017;164(9):E270–5.
[32] Boublik T. Hard-sphere equation of state. J Chem Phys 1970;53(1):471–2. https://doi.org/10.1063/1.1673824.
[33] Mansoori GA, Carnahan NF, Starling KE, Leland Jr TWL. Equilibrium thermodynamic properties of the mixture of hard spheres. J Chem Phys 1971;54(4):1523–5. https://doi.org/10.1063/1.1675048.
[34] Biesheuvel PM, Fu Y, Bazant MZ. Electrochemistry and capacitive charging of porous electrodes in asymmetric multicomponent electrolytes. Russ J Electrochem 2012;48:580–92.

[35] Bazant MZ, Kilic MS, Storey BD, Ajdari A. Towards an understanding of induced-charge electrokinetics at large applied voltages in concentrated solutions. Adv Colloid Interface Sci 2009;152(1–2):48–88.

[36] Peters PB, van Roij R, Bazant MZ, Biesheuvel PM. Analysis of electrolyte transport through charged nanopores. Phys Rev E 2016;93(5):053108.

[37] Shapovalov VL, Brezesinski G. Breakdown of the Gouy-Chapman model for highly charged Langmuir monolayers: counterion size effect. J Phys Chem B 2006;110(20):10032–40.

[38] Zhou S, Wang Z, Li B. Mean-field description of ionic size effects with nonuniform ionic sizes: a numerical approach. Phys Rev E 2011;84:021901.

[39] Xu P, Drewes JE, Heil D, Wang G. Treatment of brackish produced water using carbon aerogel-based capacitive deionization technology, Water Res 2008;42(10):2605–17. https://doi.org/10.1016/j.watres.2008.01.011. http://www.sciencedirect.com/science/article/pii/S0043135408000274.

[40] Shapira B, Avraham E, Aurbach D. The feasibility of energy extraction from acidic wastewater by capacitive mixing with a molecular-sieving carbon electrode. ChemSusChem 2016;9(24):3426–33.

[41] Dykstra JE, Dijkstra J, van der Wal A, Hamelers HVM, Porada S. On-line method to study dynamics of ion adsorption from mixtures of salts in capacitive deionization, Desalination 2016;390:47–52. https://doi.org/10.1016/j.desal.2016.04.001. http://www.sciencedirect.com/science/article/pii/S0011916416301369.

[42] López-García JJ, Horno J, Grosse C. Differential capacitance of the diffuse double layer at electrode-electrolyte interfaces considering ions as dielectric spheres: Part I. binary electrolyte solutions, J Colloid Interface Sci 2017;496:531–9. https://doi.org/10.1016/j.jcis.2017.02.043. http://www.sciencedirect.com/science/article/pii/S0021979717302151.

[43] Valette G. Double layer on silver single-crystal electrodes in contact with electrolytes having anions which present a slight specific adsorption: Part I. the (110) face. J Electroanal Chem Interfacial Electrochem 1981;122:285–97. https://doi.org/10.1016/S0022-0728(81)80159-3. http://www.sciencedirect.com/science/article/pii/S0022072881801593.

Section B

Materials

Chapter 4

Selection of Carbon Electrode Materials

Xingtao Xu*, Junfeng Li*, Yuquan Li*, Bing Ni*, Xinjuan Liu† and Likun Pan*

**Shanghai Key Laboratory of Magnetic Resonance, School of Physics and Materials Science, East China Normal University, Shanghai, People's Republic of China*

†*Institute of Coordination Bond Metrology and Engineering (CBME), China Jiliang University, Hangzhou, People's Republic of China*

4.1 Introduction

Capacitive deionization (CDI) is known as an emerging desalination technology that desalinates water by storing ions in an electrical double layer (EDL) on high-surface-area materials [1]. Since the concept of CDI was first proposed in the mid-1960s [2, 3], much effort has been devoted to developing high-surface-area materials for CDI applications. In particular, porous carbons with various dimensions have attracted increasing interest because they usually have high porosity and good electrical conductivity [4], thus showing the potential for excellent CDI performance.

Currently, several kinds of porous carbons have been studied as CDI electrodes, such as activated carbon (AC) [5, 6], carbon nanotubes (CNTs) [7, 8], carbon aerogels (CAs) [9, 10], mesoporous carbon (MC) [11, 12], graphene [13, 14], etc., but they usually suffer from poor electrosorption capacity, limiting their large-scale application. Recently, nitrogen doping has emerged as a novel strategy to functionalize porous carbons. Compared with pristine carbons, nitrogen-doped carbons usually exhibit improved wettability, electrical conductivity, and pore structure as well as provide extra pseudocapacitance [15, 16]. Therefore nitrogen-doped carbons show great promise for applications in CDI, and a great deal of effort is currently being devoted to developing nitrogen-doped carbons for such applications. As expected, such nitrogen-doped carbons show impressively improved electrosorption capacity as compared with pristine carbons.

This chapter summarizes recent progress in preparation of representative carbon materials with various dimensions, structures, and compositions and their application in CDI. Representative pure carbons and nitrogen-doped carbons are discussed in detail in this chapter.

Interface Science and Technology, Vol. 24. https://doi.org/10.1016/B978-0-12-811370-7.00004-8

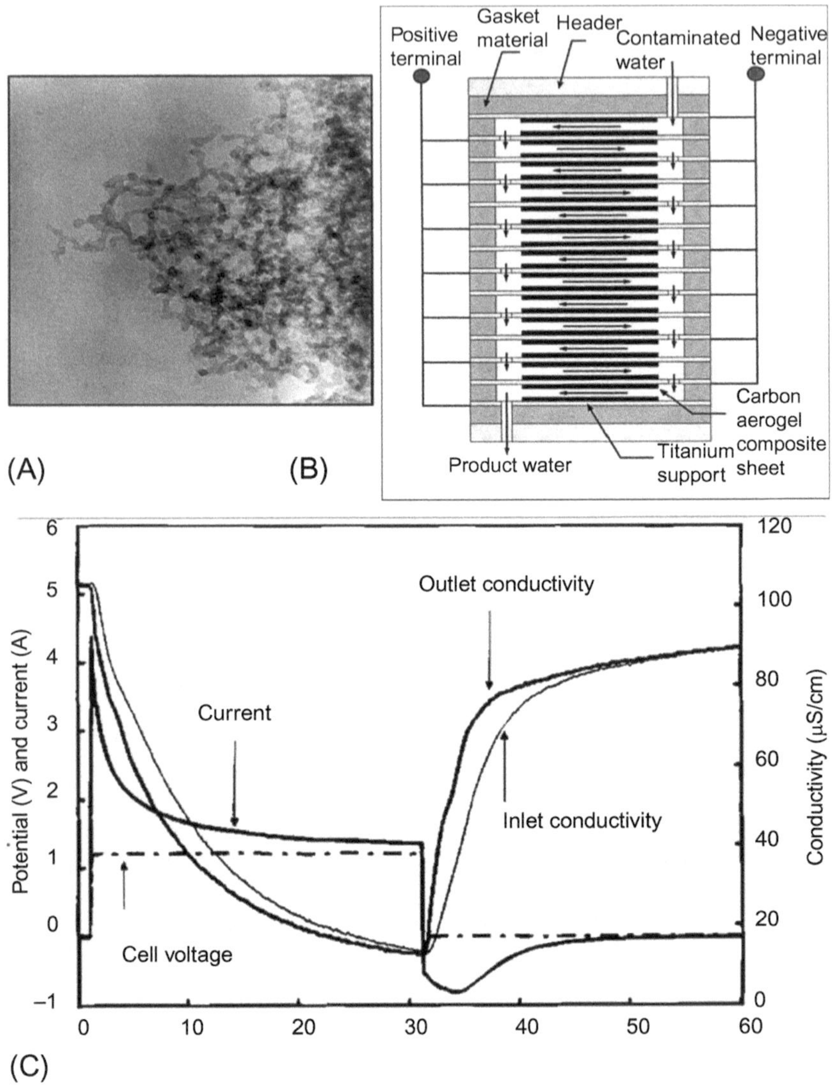

FIG. 4.1 (A) CA microstructure: transmission electron micrograph at 500,000 × magnification. (B) Schematic representation of an electrochemical adsorption-regeneration cell. (C) Typical desalination results using CA electrodes.

4.2 Carbons

4.2.1 Carbon aerogels

CAs are known as porous materials composed of interconnected and uniform carbonaceous particles (3–30 nm) with small interstitial pores (<50 nm) (Fig. 4.1A). Their continuous and monolithic structure leads to a very high

Brunauer Emmett Teller (BET) specific surface area (400–1100 m^2/g) and exceptional electrical conductivity (25–100 S/cm). Notably their chemical composition, microstructure, and physical properties can be controlled at the nanometer scale, giving rise to unique optical, thermal, acoustic, mechanical, and electrical properties. In contrast to conventional AC materials, the porosity and surface area of CAs can be controlled over a broad range by adjusting the synthesis parameters. In addition, the exceptional electrical conductivity and thin construction of CA electrodes minimize potential drop and therefore more ions can be electrosorbed on a unit surface area of CAs.

The pioneering work of employing CA electrodes in a CDI stack (Fig. 4.1B) was conducted by Farmer et al. [9]. In their work, CA film electrodes with a thickness of ~125 μm and bulk density of ~0.6 g/cm^3 were obtained by infiltrating 70% w/v resorcinol-formaldehyde solution into a porous carbon paper and then pyrolyzing at 1050°C in nitrogen atmosphere. Fig. 4.1C shows the typical desalination results using CA electrodes. As compared with the conventional AC-based CDI electrodes, the CA electrodes show an anion capacity as good as those of the best ACs (highest surface area), due to the fact that the resistance of a CA electrode is much less than that of a comparable AC electrode. On the other hand, a large fraction of micropores (especially with a diameter less than 1 nm) in ACs plays a negative role in contributing to the electrosorption capacity, although they have contributed greatly to BET surface area. Beyond this conclusion, several basic questions, including parasitic reactions, system reliability, chromatographic separation of ions, and ionic selectivity, which were not involved in previous works [9, 17, 18], have also been pointed out. These issues associated with CA-based CDI electrodes are further addressed in some studies, discussed in the following paragraphs.

Later on, the most comprehensive study of the application of CAs in CDI was reported by Gabelich et al. [19]. They conducted a thoughtful study to elucidate the effect of ionic charge, ionic radius, pH, and mass on the electrosorption performance of CAs. A laboratory-scale CDI electrode stack containing CAs was prepared under three different conditions and used to separate salts from artificial single and multisalt solutions. In addition, streams with more complex compositions, including an unfiltered blend of 75% Colorado River water and 25% water from the California State Water Project, with/without small amounts of organic material, were studied so as to check the impact of natural organic matter on CDI performance. The results showed that monovalent ions with smaller hydrated radii were removed from the solutions preferentially over multivalent ions. Two important findings emerged from this study: one was that only a few percent (14–42 m^2/g) of the BET surface area of carbon aerogels (400–590 m^2/g) are available for electrosorption; another was that the presence of organic matter in the treated water significantly reduced electrosorption capacity of ions due to fouling of the carbon aerogel surface and perhaps also its pore system.

Many researchers have devoted major efforts to the improvement of desalination performance by means of optimization of the properties of as-prepared CAs. Jung et al. [20] explored the CDI performance of nanostructured CA electrodes synthesized via ambient drying. The activated CA electrodes retain high porosity (~80%), low bulk density (~0.50 g/cm^3), high specific surface area (~610 m^2/g), high electrical conductivity (~13.2 S/cm), and high specific capacitance (~220 F/g). Furthermore, a high ion removal efficiency of 97.6% was also achieved in a 50 mg/L NaCl solution with a flow rate of 400 mL/min and an applied voltage of 1.7 V.

The long-term stability of CDI cells is another important parameter for evaluating the CDI electrodes. Gao et al. [21] investigated the reasons for degeneration of CDI cells by performing a long-term research study using a flow-through CDI cell assembled with CA electrodes. The results showed that the potential of zero charge (PZC) location for the anodic shifting to more positive values could be the main reason for the degeneration of CDI cells. The pristine electrode and used anode are presented in Fig. 4.2A and B. Further experiments proved that PZC shifting is caused by the anodic oxidation of the electrode in aqueous solution (Fig. 4.2C).

4.2.2 CNTs

Since the first discovery of CNTs by Iijima in 1991, CNTs have attracted enormous attention for their fundamental behavior and for their use in a wide variety of applications in nanodevices. CNTs have unique mechanical, electrical,

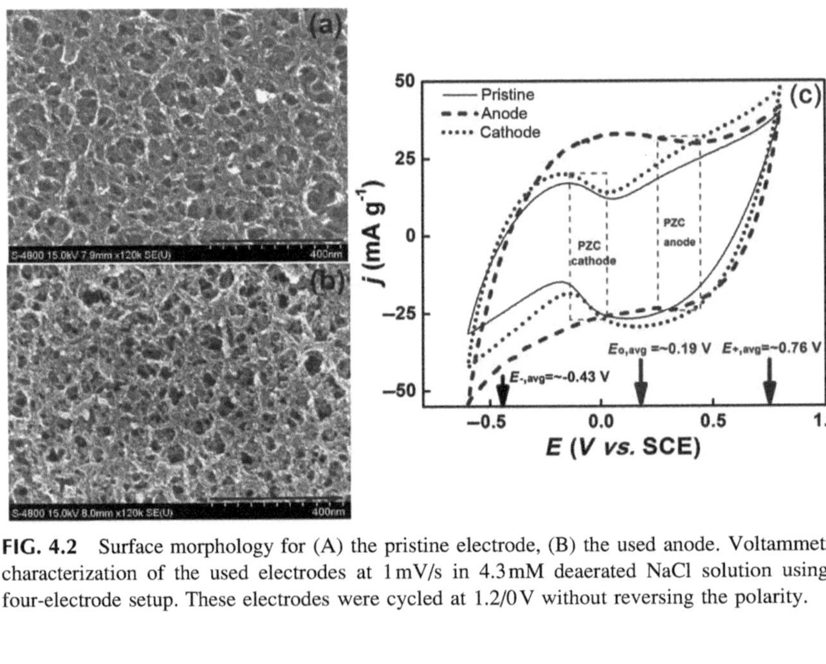

FIG. 4.2 Surface morphology for (A) the pristine electrode, (B) the used anode. Voltammetric characterization of the used electrodes at 1 mV/s in 4.3 mM deaerated NaCl solution using a four-electrode setup. These electrodes were cycled at 1.2/0 V without reversing the polarity.

magnetic, optical, and thermal properties, which make them an excellent candidate for application in nanoelectronics, nanosensors, and supercapacitors, as well as in CDI.

CNTs can be synthesized by different techniques, including arc-discharge, laser ablation, and various catalytic chemical vapor depositions (CCVDs). Typical scanning electronic microscopy (SEM) and TEM images of CNTs presented in Fig. 4.3A and B [22] show the curled structure of individual nanotubes. The nanotubes are highly entangled with each other as a result of van der Waals forces of attraction and form an interconnecting structure.

Many methods for treating CNTs before fabricating CDI electrodes have been summarized in Table 4.1. Generally, the as-synthesized CNTs are first refluxed in concentrated acid in order to remove residual catalyst derived from the CVD process and introduce various functional groups into CNTs,

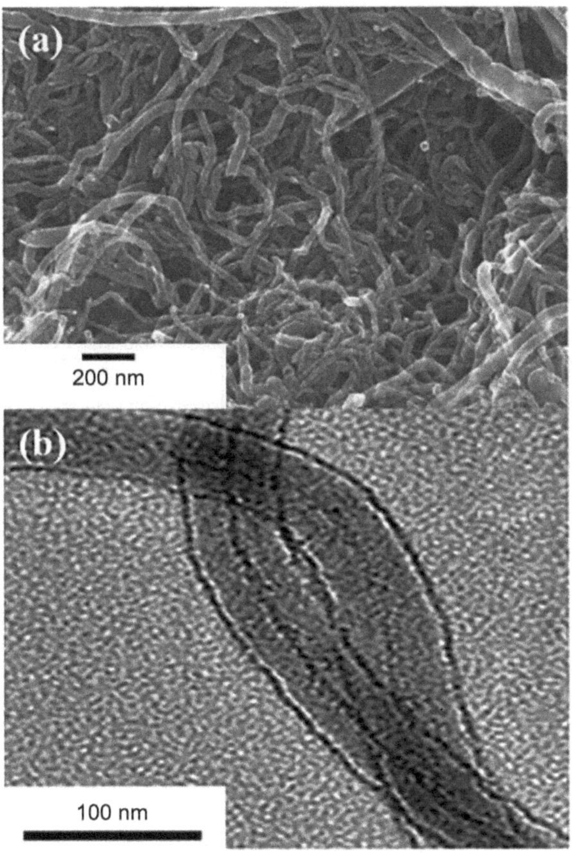

FIG. 4.3 Typical (A) SEM and (B) TEM images of CNTs.

TABLE 4.1 The Effect of Various Treatments on Porous Properties of CNTs

CNTs-Treated Methods	BET Surface Area (m^2/g)	Microporous Volume (cm^3/g)	Mesoporous Volume (cm^3/g)
As-synthesized	150–1587	0.06–0.15	0.85
HNO_3 treated	157	–	0.37
NH_3 treated	195	–	0.42
KOH treated	785	0.17	1.04
Air activated	270	0.06	0.56
CO_2 activated	420	0.1	0.67
Ozone treated	320	0.12	0.69
Heat treatment	550	0.18	0.97

such as carboxyl and hydroxyl groups, which can improve the hydrophilic properties and BET surface area of CNTs. Then the mixture of CNTs and binder materials in a certain weight ratio is pressed under 10–30 MPa pressure for several minutes, further carbonized at a high temperature (800–1000°C) and finally cooled down for electrode fabrication.

Shi's group conducted a series of electrosorption experiments by using multiwalled CNTs with binding agent as CDI electrodes [23]. Their studies showed that pretreated CNT electrodes after carbonization show the largest specific surface area and specific pore volume as well as exhibit a good NaCl electrosorption performance. The percentage of NaCl desorption was nearly 90% when the initial NaCl concentration, voltage, water flow rate, and number of electrode pairs were 5000 mg/L, 1.2 V, 10 mL/min, and 40, respectively. In addition, the regeneration process could be carried out easily within a short period of time.

4.2.3 Mesoporous carbon

According to the International Union of Pure and Applied Chemistry (IUPAC) recommendations, porous carbon materials can be classified into three types based on their pore radii: micropore <2 nm, 2 nm $<$ mesopore $<$ 50 nm, and macropores $>$ 50 nm. Traditional porous carbon materials, such as AC and carbon molecular sieves, are synthesized by pyrolysis and physical or chemical activation of organic precursors. These carbon materials normally have relatively broad and disordered pore distribution in both micropore and mesopore ranges. As reported elsewhere [24], the effect of EDL overlapping within microporous carbons is significant and must be considered, that is, micropores with diameter smaller than a cutoff pore width do not contribute to electrosorption. Taking this point into consideration, porous carbons with dominant and ordered mesopores would be more suitable for CDI.

Conventional synthetic methods for ordered mesoporous carbons (OMCs) can be simply classified as either hard-template or soft-template synthesis. The former approach [25] includes the following steps that have been extensively used: (a) preparation of silica gel with a controlled pore structure, (b) impregnation/infiltration of the silica template with monomer or polymer precursors, (c) cross-linking and carbonization of the organic precursors, and (d) dissolution of the silica template. A typical TEM image of ordered carbon molecular sieve is displayed in Fig. 4.4, showing an ordered porous structure accordingly. Amphiphilic molecules, such as surfactants and block copolymers, have been extensively employed as soft templates in the synthesis of OMCs. Compared with the hard-template method, the use of mesoporous polymers as soft-template allows the efficient fabrication of MCs with controlled pore structure via simply carbonizing the polymer template. However, the synthesis of OMCs via the self-assembly approach is still difficult to achieve.

Zou's group performed a series of CDI experiments by employing various OMCs as electrodes [26, 27]. Compared with ACs, the ion amount adsorbed by OMCs was higher than that by ACs (11.6 μmol/g for OMCs and 4.3 μmol/g for ACs) when the initial concentration of NaCl solution was 25.6 mg/L. The preferable electrosorption performance of OMC electrodes might be attributed

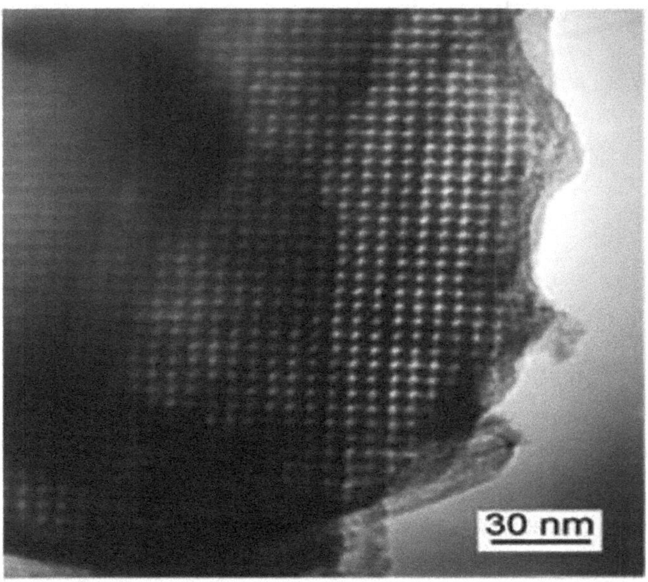

FIG. 4.4 TEM image of the ordered carbon molecular sieve.

to their optimal pore radius (average 3.3 nm) and ordered mesoporous structure, which could provide a fast diffusion path for salt ions during the adsorption and desorption process. Conversely, for the AC electrodes, some surface area appears to be unavailable for ion adsorption due to the fact that some micropores are too small for the ions to access and an EDL overlapping effect becomes obvious [27]. Zou's group further studied the electrosorption performance of various OMC samples with the addition of nickel salts in a sol-gel process. Fig. 4.5A–C show the TEM images of various OMC samples. In their work, the products prepared by the addition of $NiSO_4{\cdot}6H_2O$ and $Ni(NO_3)_2{\cdot}6H_2O$ were denoted as OMC-S and OMC-N, respectively. OMC without the use of nickel salt was prepared by fixing the mass ratio of other raw materials, and was denoted as OMC-W. The results showed that OMC-S exhibited the best electrochemical performance, with an electrosorption capacity of 15.9 μmol/g (Fig. 4.5D) and good regeneration ability (Fig. 4.5E) as compared with other OMCs (more detailed information can be found from the literature [27]) and a commercial AC. Such superior desalination performance could be easily attributed to its high specific surface area (1491 m^2/g) and ordered mesopores (3.7 nm).

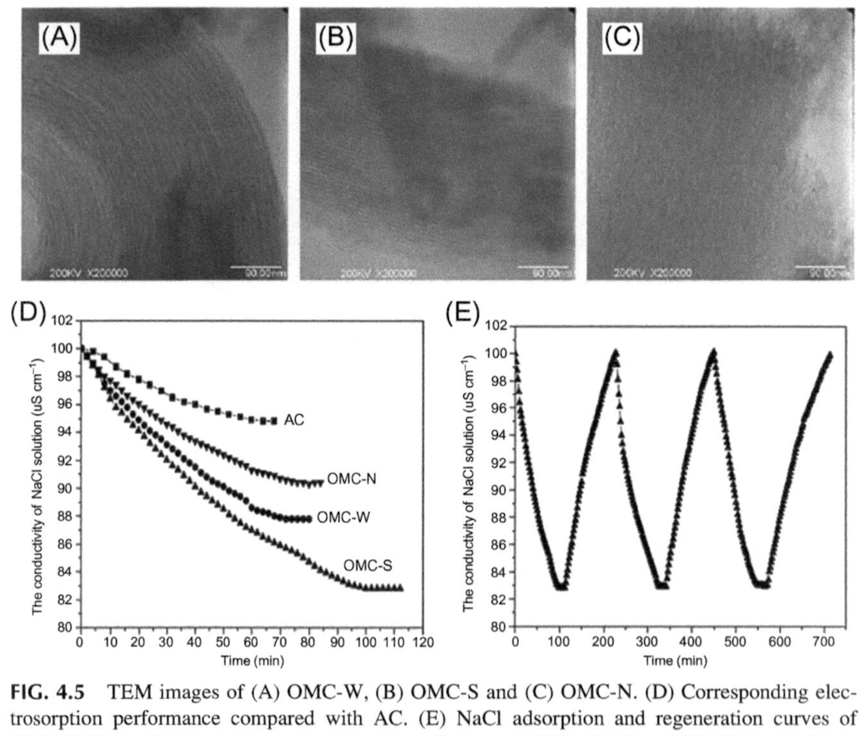

FIG. 4.5 TEM images of (A) OMC-W, (B) OMC-S and (C) OMC-N. (D) Corresponding electrosorption performance compared with AC. (E) NaCl adsorption and regeneration curves of OMC-S. Note: The products prepared by the addition of $NiSO_4{\cdot}6H_2O$ and $Ni(NO_3)_2{\cdot}6H_2O$ are denoted as OMC-S and OMC-N, respectively, where S and N indicate the kind of nickel salt used. OMC without nickel salt is prepared by fixing the mass ratio of other raw materials, and is denoted as OMC-W, where W represents the absence of nickel salt.

Furthermore, Dai's group [12] reported a method to prepare activated MC with a ultrahigh surface area for the electrosorption of ions from aqueous solutions. Through optimizing the activation conditions, the surface area of activated MC could reach as high as $2000\,m^2/g$ with large pore volumes (up to $\sim 1.6\,cm^3/g$) but a highly mesoporous structure was still maintained. Compared with pristine MC, the electrosorption capacity of activated MC was greatly enhanced.

4.2.4 Graphene

Graphene is known as a two-dimensional (2D) structured carbon material with novel physical and chemical properties that has aroused extensive attention in the field of new technologies, including energy storage, biosensors, and optoelectronics. Theoretically, the surface area of a single graphene layer can reach as high as $2630\,m^2/g$ and the corresponding conductivity is as high as 7200 S/m, substantially higher than that of AC. Several studies [28] have successfully applied graphene as an electrode material in supercapacitors, and achieved a high specific capacitance of more than 200 F/g in aqueous electrolytes. Such an excellent capacitive performance of graphene also enables it to be a promising electrode material in CDI.

Pan's group first introduced graphene into CDI in 2009 [29]. The graphene was synthesized by a modified Hummer's method and hydrazine reduction and then dip-coated on a graphite layer as a CDI electrode. As shown in Fig. 4.6A and B, the SEM and TEM images of graphene showed that the corrugated and scrolled sheets resemble a flower shape and porous structure, which facilitates the high storage capacity for ions from solution. However, the specific surface area of as-prepared graphene in this work was very low (about $14\,m^2/g$), indicating the incomplete exfoliation and the serious aggregation between each layer due to the presence of van der Waals forces. Further studies were carried out to optimize and improve the preparation to address this issue. By changing the ratio of sulfuric acid and nitric acid in the pretreatment, the specific surface area of graphene could reach as high as $222.01\,m^2/g$ and the specific electrosorption capacity was 23.18 μmol/g when the initial concentration of NaCl was 25 mg/L (Fig. 4.6C), which was higher than that of AC under the same experimental conditions [30]. Furthermore, the selective electrosorption experiment (Fig. 4.6D) conducted in solutions of NaCl, $CaCl_2$, $MgCl_2$, and $FeCl_3$ showed similar results to carbon aerogel and CNT-based CDI electrodes [31].

4.3 Nitrogen-Doped Carbons

4.3.1 Nitrogen-doped CAs

CAs are one of the earliest developed materials in CDI [24, 32, 33]. They usually show a continuous, monolithic structure with a high specific surface area (400–$1100\,m^2/g$) and good electrical conductivity (25–100 S/cm) [34, 35],

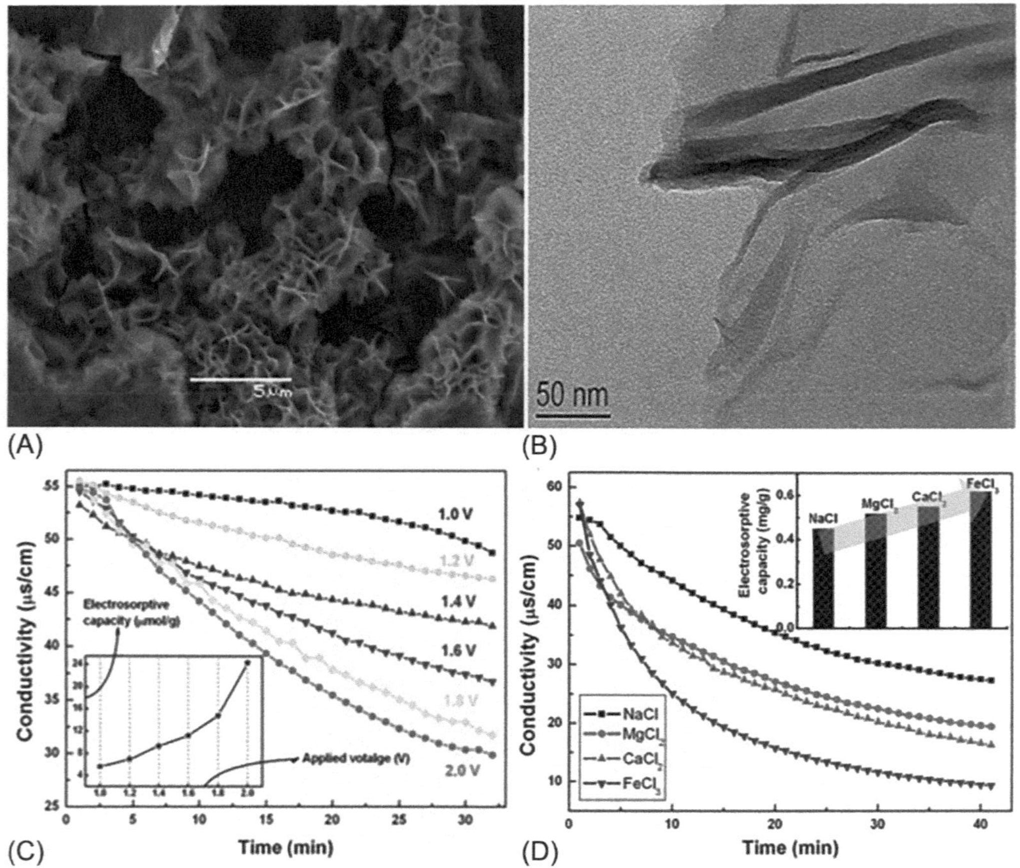

FIG. 4.6 (A) SEM and (B) TEM images of graphene; (C) CDI performance of graphene at different working voltages; (D) electrosorption capacities of graphene in the solutions with different ion species. The applied electrical voltage and initial concentration was 2.0 V and 50 μS/cm and the feed solutions were NaCl, $CaCl_2$, $MgCl_2$, and $FeCl_3$, respectively.

which makes a significant contribution to their electrosorption capacity. Gloria Rasines et al. [35] first introduced nitrogen-doped CAs into CDI in 2014. The hydrogels were synthesized by the polycondensation of melamine (M), resorcinol (R) and formaldehyde (F) using sodium carbonate as catalyst (C) and deionized water (W) as solvent. Then the hydrogels were dried under CO_2 supercritical conditions and carbonized at 750°C under nitrogen atmosphere and nitrogen-doped CAs. Finally, the nitrogen-doped CAs were further activated using CO_2 at 750°C. The samples were denoted as MRF(n)-X-Y, where $n=0$ or 1 indicating nonprepolymerized and prepolymerized aerogels; X accounts for the M/R molar ratio and Y for the pH value.

Fig. 4.7 shows the SEM images of MRF(0)-0.5-6 and MRF(1)-0.5-6. It can be seen that MRF(0)-0.5-6 is formed by large spherical particles, as opposed to the more compact interconnected smaller particles for MRF(1)-0.5-6. The CDI experiments were conducted in a continuously recycling system at 0.9 V, 1.2 V, or 1.5 V using 146.1 and 584.4 mg/L NaCl solution as feed solution. The total solution volume was 20 mL. Fig. 4.8A and B describes the electrosorption behaviors of MRF(0)-0.5-6 and MRF(1)-0.5-6 electrodes. The electrosorption capacities of 8.2 and 5.4 mg/g were measured for MRF(0)-0.5-6 and MRF(1)-0.5, respectively, at 1.5 V and 0.025 M. Notably, these adsorption values were higher than those reported for nondoped aerogels subjected to similar conditions, evidencing the positive effect of doping on the electrochemical behavior [33].

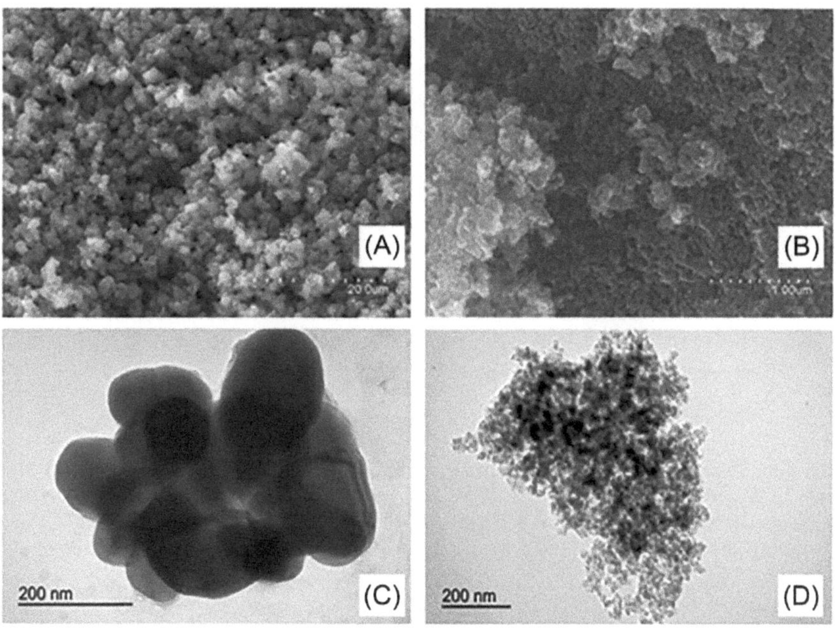

FIG. 4.7 TEM images of activated NCAs, (A) MRF(0)-0.5-6, (B) MRF(1)-0.5-6, (C) MRF(0)-0.5-8, (D) MRF(1)-0.5-8, (E) MRF(0)-2-8, (F) MRF(1)-2-8. SEM micrographs of (H) MRF(0)-0.5-6, (H) MRF(1)-0.5-6.

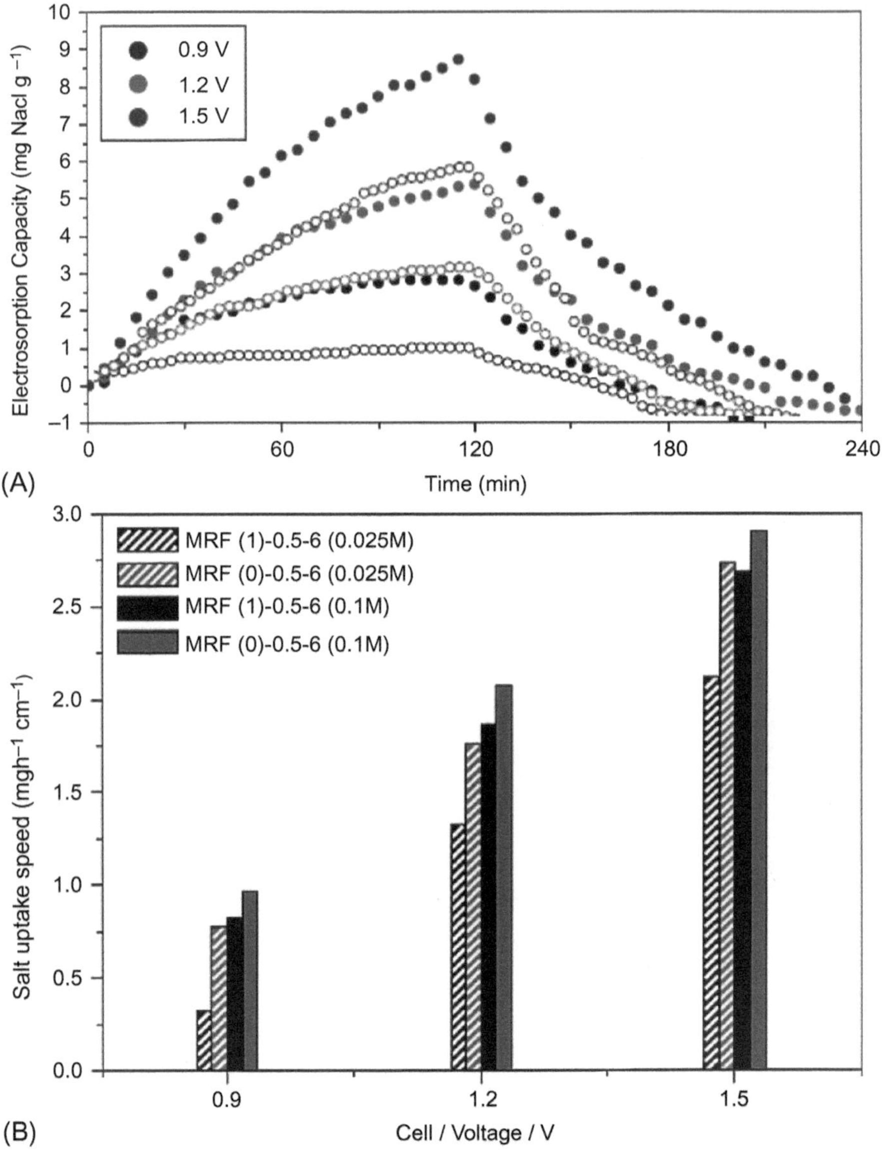

FIG. 4.8 Capacitive deionization experiments performed at different potentials in 0.025 M NaCl solution: (A) Electrosorption capacity of MRF(0)-0.5-6 (closed symbols) and MRF(1)-0.5-6 (open symbols); (B) rate of salt uptake at 0.025 and 0.1 M NaCl at an applied voltage of 0.9, 1.2 and 1.5 V for MRF(0)-0.5-6 and MRF(1)-0.5-6. (A color version of this figure can be viewed online.).

4.3.2 Nitrogen-doped graphene

As a 2D carbon nanostructure with novel physical and chemical properties, graphene has drawn extensive attention in many fields since 2004 [36–39]. Chemical approaches, via reduction of graphite oxide (GO), for the large-scale production of reduced graphene (RG) sheets have been realized [40]. In theory, the surface area of a single graphene layer can be as high as 2630 m^2/g and the corresponding conductivity is 7200 S/m, much higher than that of AC. These favorable properties endow graphene with the ability to be used as electrode material for energy storage devices and CDI [41, 42]. Pan et al. [43] first introduced RG into CDI in 2009. The RG was synthesized by a modified Hummer's method and hydrazine reduction. The electrosorption capacity of RG was calculated to be ~1.4–1.9 mg/g when the initial concentration of NaCl was 25 mg/L. However, the obtained electrosorption performance of pristine graphene (PG) is actually much lower than the anticipated estimated value [44], which limits its application in CDI. Therefore, the same group later developed nitrogen-doped graphene (NG) via a simple thermal treatment of graphene oxide in an ammonia atmosphere at 800°C with a rate of 10°C/min.

As shown in Fig. 4.9A, the NG synthesized with NH_3 shows a loosely layered structure and the gap between the sheets is obvious due to the effective exfoliation. Because the 2D membrane structure is unstable in thermodynamics, NG is naturally corrugated and scrolled, and the pores are formed between the lamellar sheets. Such a porous structure is clearly beneficial for facilitating the high storage capacity for ions from solution. The CDI experiments were conducted in a continuously recycling system using 50 mg/L NaCl solution as feed solution. The total solution volume was 50 mL. Fig. 4.9B describes the electrosorption behaviors of NG electrodes. It can be seen that higher voltage leads to a higher electrosorption capacity because of the stronger electrostatic force. The electrosorption capacities of NG increase from 1.29 to 4.81 mg/g at an applied voltage from 1.2 to 1.8 V, much higher than those of PG (from 0.96 to 3.85 mg/g). The enhanced electrosorption performance of NG is ascribed to

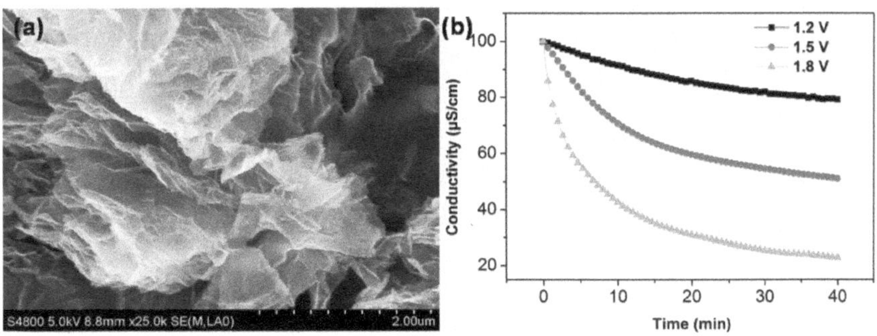

FIG. 4.9 SEM images of PG (A) and NG (B). Electrosorption performance of (A) NG and (C) PG.

higher specific surface area and pore volume, lower charge transfer resistance, and enhanced specific capacitance by nitrogen doping.

Creating three-dimensional (3D) interconnected carbon structures has been currently a hotspot in the CDI field. Pan et al. first introduced graphene sponge (GS) into CDI in 2015. The GS was fabricated via directly freeze-drying a graphene oxide solution followed by annealing in a nitrogen atmosphere. The electrosorption capacity of GS reaches as high as 14.9 mg/g when the initial concentration of NaCl was 500 mg/L. There is no denying the fact that these 3D GSs can prevent the agglomeration of graphene sheets, enlarge the specific surface area, and ultimately improve the electrosorption capacity up to 15.1 mg/g. However, further enhancement of electrosorption capacity is still necessary to meet the practical demands in desalination applications. Undoubtedly, nitrogen doping should be a promising solution.

The same group developed nitrogen-doped GS (NGS) via a facile and low-cost strategy [45], as shown in Fig. 4.10A. Graphene oxide sponge (GOS) was produced by a freeze-drying process. After annealing in NH_3, NGS was obtained. As shown in Fig. 4.10B, NGS exhibits an interconnected porous 3D framework of randomly corrugated and scrolled graphene nanosheets. These curls and

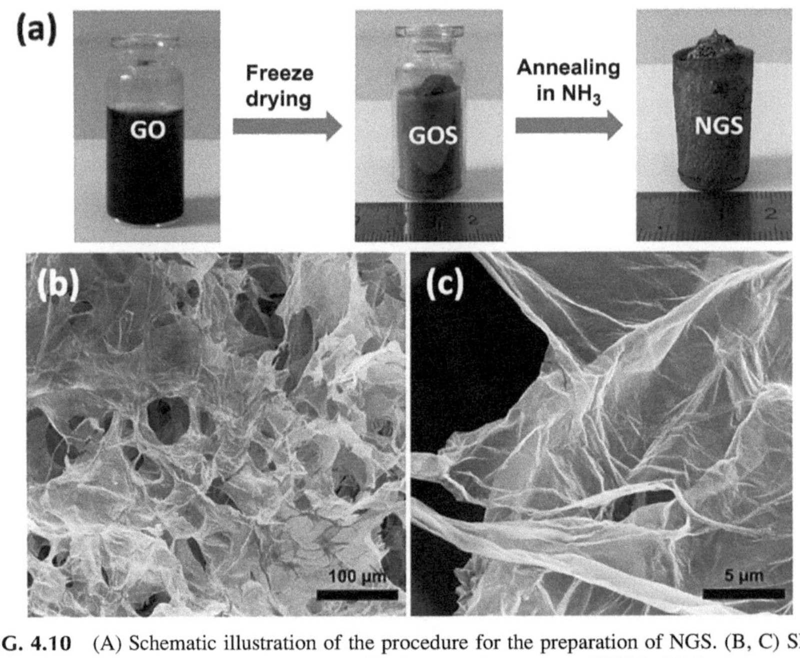

FIG. 4.10 (A) Schematic illustration of the procedure for the preparation of NGS. (B, C) SEM images showing the interconnected, porous 3D framework structure of NGS. (B) Interconnecting porous network between the graphene layers. Scale bar: 100 μm. (C) Electrosorption capacity for NGS, GS, and PG electrodes over 30 min in NaCl solution with an initial concentration of ~50 mg/L at an applied voltage of 1.5 V.

wrinkles could prevent graphene sheets from restacking with each other. The CDI experiments were conducted in a continuously recycling system at 1.2V in NaCl solution with a concentration of 500mg/L (Fig. 4.10C). The total solution volume was 50mL. Fig. 4.10C describes the electrosorption behaviors of PG, GS, and NGS electrodes. The electrosorption capacity of NGS is calculated to be 21.0mg/g, which is about 1.4 and 4.6 times those of GS (14.6mg/g) and PG (4.5mg/g), respectively. The ultrahigh electrosorption capacity of NGS can be ascribed to various reasons: first of all, the 3D porous structure of NGS with an enhanced specific surface area (526.7 m^2/g) and an increased pore volume (3.13 cm^3/g) provides short ion diffusion pathways and more available spaces to accommodate ions during the electrosorption. Secondly, nitrogen doping increases the pseudocapacitance and improves the electrical conductivity of graphene and the wettability of the interface between the electrolyte and the electrodes, which can decrease the inner resistance of electrodes and help to improve the electrosorption performance of NGS.

4.3.3 Nitrogen-doped porous carbon spheres

Spherical structures have been demonstrated to be an attractive morphology in energy storage fields like supercapacitor [46, 47] and secondary battery [48, 49], since this unique structure enjoys a high accessible surface area, low resistance, maximal structural stability, excellent chemical stability, and convenience for preparing electrodes. Considering these points, disordered carbon with a spherical structure should have the potential to be an excellent anode material for CDI. Pan et al. introduced porous carbon spheres (PCSs) into CDI in 2014. The PCSs were fabricated through a fast microwave-assisted approach using sucrose as the precursor in a microwave system and subsequent carbonization. The electrosorption capacities of PCSs was calculated to be 5.81mg/g when the initial concentration of NaCl was 500mg/L. However, current performance of PCSs is still needed to be improved in order to meet the practical demand. Therefore, the same group [50] developed nitrogen-doped PCS (NPCSs) through a microwave-assisted approach and subsequent thermal treatment in NH_3 atmosphere. Fig. 4.11A depicts the SEM image of NPCSs. It can be seen that a spherical structure is obtained though microwave-assisted reaction, and the NPCSs with porous surfaces have an average diameter of around 800nm. Obviously, the spherical structure remains intact after the nitrogen doping process. The CDI experiments were conducted in a continuously recycling system at 1.2V using NaCl solution as feed solution. The total solution volume was 50mL. Fig. 4.11B describes the electrosorption performance of PCSs. It can be seen that the electrosorption capacity increases with the increase in the NaCl concentration and reaches a highest value of 14.91mg/g. Compare with PCSs, as shown in Fig. 4.11C, the electrosorption capacities of NPCSs have been doubled in both concentrations, which suggests that the nitrogen doping is a very effective method to improve the capacitive and electrosorption performance.

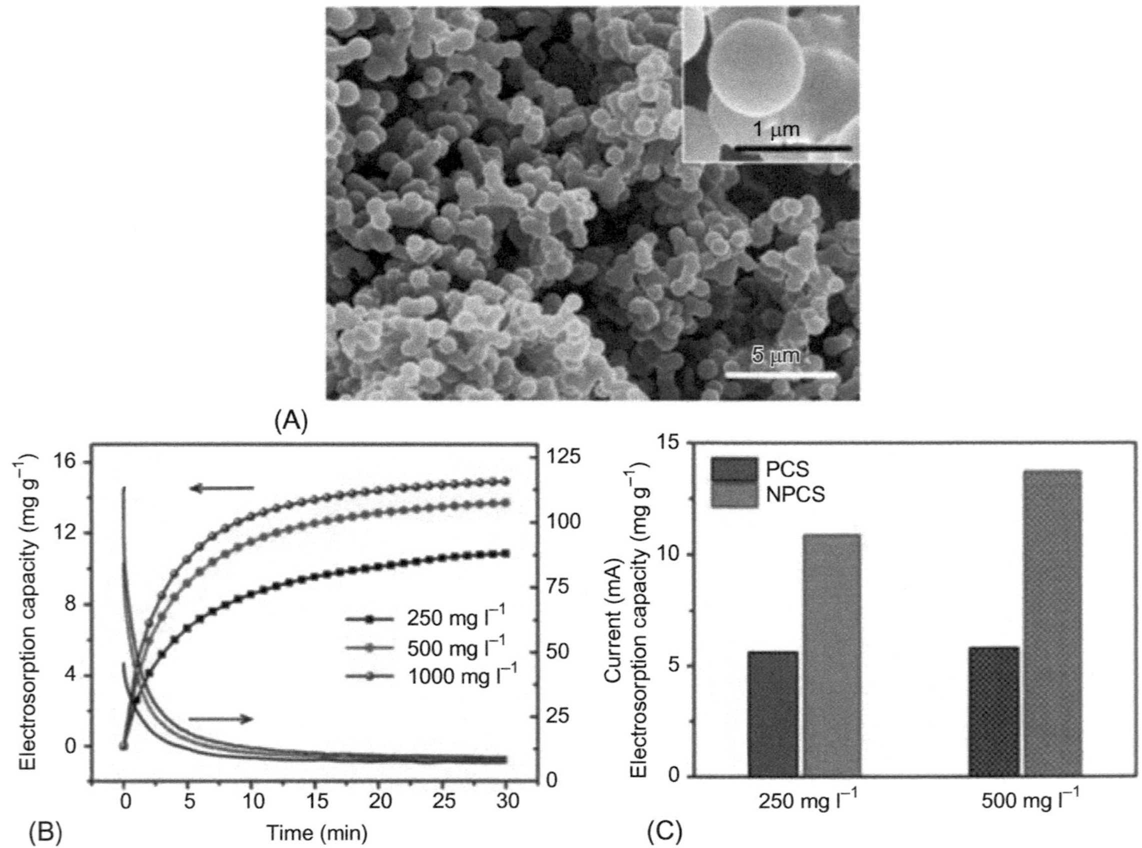

FIG. 4.11 (A) SEM image of NPCSs. Inset shows the high-magnification image. (B) Electrosorption capacity and current transient for NPCSs electrode over 30 min in NaCl solution with different initial concentrations. (C) Electrosorption capacity comparisons of PCSs and NPCSs in NaCl solution with different initial concentrations.

In summary, porous carbons with diverse dimensions, structures, and compositions have been prepared and used in CDI applications in recent decades. Currently, designing nanostructured carbons and nitrogen-doped carbons are hotspots and it is also highly desirable to expand new carbon families and further explore their potential for CDI applications.

References

[1] Oren Y. Capacitive deionization (CDI) for desalination and water treatment—past, present and future (a review). Desalination 2008;228:10–29.
[2] Murphy G, Caudle D. Mathematical theory of electrochemical demineralization in flowing systems. Electrochim Acta 1967;12:1655–64.
[3] Johnson A, Newman J. Desalting by means of porous carbon electrodes. J Electrochem Soc 1971;118:510–7.
[4] Liu Y, Nie C, Liu X, Xu X, Sun Z, Pan L. Review on carbon-based composite materials for capacitive deionization. RSC Adv 2015;5:15205–25.
[5] Zou L, Morris G, Qi D. Using activated carbon electrode in electrosorptive deionisation of brackish water. Desalination 2008;225:329–40.
[6] Mossad M, Zou L. A study of the capacitive deionisation performance under various operational conditions. J Hazard Mater 2012;213:491–7.
[7] Wang L, Wang M, Huang Z-H, Cui T, Gui X, Kang F, Wang K, Wu D. Capacitive deionization of NaCl solutions using carbon nanotube sponge electrodes. J Mater Chem 2011;21:18295–9.
[8] Nie C, Pan L, Li H, Chen T, Lu T, Sun Z. Electrophoretic deposition of carbon nanotubes film electrodes for capacitive deionization. J Electroanal Chem 2012;666:85–8.
[9] Farmer JC, Fix DV, Mack GV, Pekala RW, Poco JF. Capacitive deionization of NaCl and $NaNO_3$ solutions with carbon aerogel electrodes. J Electrochem Soc 1996;143:159–69.
[10] Xu P, Drewes JE, Heil D, Wang G. Treatment of brackish produced water using carbon aerogel-based capacitive deionization technology. Water Res 2008;42:2605–17.
[11] Tsouris C, Mayes R, Kiggans J, Sharma K, Yiacoumi S, DePaoli D, Dai S. Mesoporous carbon for capacitive deionization of saline water. Environ Sci Technol 2011;45:10243–9.
[12] Mayes RT, Tsouris C, Kiggans Jr JO, Mahurin SM, DePaoli DW, Dai S. Hierarchical ordered mesoporous carbon from phloroglucinol-glyoxal and its application in capacitive deionization of brackish water. J Mater Chem 2010;20:8674–8.
[13] Xu X, Pan L, Liu Y, Lu T, Sun Z. Enhanced capacitive deionization performance of graphene by nitrogen doping. J Colloid Interface Sci 2015;445:143–50.
[14] Xu X, Pan L, Liu Y, Lu T, Sun Z, Chua DH. Facile synthesis of novel graphene sponge for high performance capacitive deionization. Sci Rep 2015;5:srep08458.
[15] Wen Z, Wang X, Mao S, Bo Z, Kim H, Cui S, Lu G, Feng X, Chen J. Crumpled nitrogen-doped graphene nanosheets with ultrahigh pore volume for high-performance supercapacitor. Adv Mater 2012;24:5610–6.
[16] Chen L-F, Zhang X-D, Liang H-W, Kong M, Guan Q-F, Chen P, Wu Z-Y, Yu S-H. Synthesis of nitrogen-doped porous carbon nanofibers as an efficient electrode material for supercapacitors. ACS Nano 2012;6:7092–102.
[17] Farmer J, Fix D, Mack G, Pekala R, Poco J. Capacitive deionization of NH_4ClO_4 solutions with carbon aerogel electrodes. J Appl Electrochem 1996;26:1007–18.

[18] Farmer JC, Bahowick SM, Harrar JE, Fix DV, Martinelli RE, Vu AK, Carroll KL. Electrosorption of chromium ions on carbon aerogel electrodes as a means of remediating ground water. Energy Fuel 1997;11:337–47.
[19] Gabelich CJ, Tran TD, Suffet IM. Electrosorption of inorganic salts from aqueous solution using carbon aerogels. Environ Sci Technol 2002;36:3010–9.
[20] Jung H-H, Hwang S-W, Hyun S-H, Lee K-H, Kim G-T. Capacitive deionization characteristics of nanostructured carbon aerogel electrodes synthesized via ambient drying. Desalination 2007;216:377–85.
[21] Gao X, Omosebi A, Landon J, Liu K. Dependence of the capacitive deionization performance on potential of zero charge shifting of carbon xerogel electrodes during long-term operation. J Electrochem Soc 2014;161:E159–66.
[22] Wang X, Li M, Chen Y, Cheng R, Huang S, Pan L, Sun Z. Electrosorption of ions from aqueous solutions with carbon nanotubes and nanofibers composite film electrodes. Appl Phys Lett 2006;89:053127.
[23] Dai K, Shi L, Fang J, Zhang D, Yu B. NaCl adsorption in multi-walled carbon nanotubes. Mater Lett 2005;59:1989–92.
[24] Ying T-Y, Yang K-L, Yiacoumi S, Tsouris C. Electrosorption of ions from aqueous solutions by nanostructured carbon aerogel. J Colloid Interface Sci 2002;250:18–27.
[25] Liang C, Li Z, Dai S. Mesoporous carbon materials: synthesis and modification. Angew Chem Int Ed 2008;47:3696–717.
[26] Zou L, Li L, Song H, Morris G. Using mesoporous carbon electrodes for brackish water desalination. Water Res 2008;42:2340–8.
[27] Li L, Zou L, Song H, Morris G. Ordered mesoporous carbons synthesized by a modified sol–gel process for electrosorptive removal of sodium chloride. Carbon 2009;47:775–81.
[28] Lv W, Tang D-M, He Y-B, You C-H, Shi Z-Q, Chen X-C, Chen C-M, Hou P-X, Liu C, Yang Q-H. Low-temperature exfoliated graphenes: vacuum-promoted exfoliation and electrochemical energy storage. ACS Nano 2009;3:3730–6.
[29] Li H, Lu T, Pan L, Zhang Y, Sun Z. Electrosorption behavior of graphene in NaCl solutions. J Mater Chem 2009;19:6773–9.
[30] Li H, Zou L, Pan L, Sun Z. Novel graphene-like electrodes for capacitive deionization. Environ Sci Technol 2010;44:8692–7.
[31] Li H, Zou L, Pan L, Sun Z. Using graphene nano-flakes as electrodes to remove ferric ions by capacitive deionization. Sep Purif Technol 2010;75:8–14.
[32] Porada S, Zhao R, Van Der Wal A, Presser V, Biesheuvel PM. Review on the science and technology of water desalination by capacitive deionization. Prog Mater Sci 2013;58: 1388–442.
[33] Rasines G, Lavela P, Macías C, Haro M, Ania CO, Tirado JL. Electrochemical response of carbon aerogel electrodes in saline water. J Electroanal Chem 2012;671:92–8.
[34] Rasines G, Lavela P, Macías C, Zafra MC, Tirado JL, Ania CO. On the use of carbon black loaded nitrogen-doped carbon aerogel for the electrosorption of sodium chloride from saline water. Electrochim Acta 2015;170:154–63.
[35] Rasines G, Lavela P, Macías C, Zafra MC, Tirado JL, Parra JB, Ania CO. N-doped monolithic carbon aerogel electrodes with optimized features for the electrosorption of ions. Carbon 2015;83:262–74.
[36] Novoselov K, Geim AK, Morozov S, Jiang D, Zhang Y, Dubonos S, Grigorieva I, Firsov A. Electric field effect in atomically thin carbon films. Science 2004;306:666–9.
[37] Geim AK, Novoselov KS. The rise of graphene. Nat Mater 2007;6:183–91.

[38] Cohen-Tanugi D, Grossman JC. Water desalination across nanoporous graphene. Nano Lett 2012;12:3602–8.
[39] Zhu Y, Murali S, Stoller MD, Ganesh K, Cai W, Ferreira PJ, Pirkle A, Wallace RM, Cychosz KA, Thommes M. Carbon-based supercapacitors produced by activation of graphene. Science 2011;332:1537–41.
[40] Choucair M, Thordarson P, Stride JA. Gram-scale production of graphene based on solvothermal synthesis and sonication. Nat Nanotechnol 2008;4:30–3.
[41] Jia B, Zou L. Graphene nanosheets reduced by a multi-step process as high-performance electrode material for capacitive deionisation. Carbon 2012;50:2315–21.
[42] Wang H, Zhang D, Yan T, Wen X, Shi L, Zhang J. Graphene prepared via a novel pyridine-thermal strategy for capacitive deionization. J Mater Chem 2012;22:23745–8.
[43] Li HB, Lu T, Pan LK, Zhang YP, Sun Z. Electrosorption behavior of graphene in NaCl solutions. J Mater Chem 2009;19:6773–9.
[44] Li HB, Pan LK, Lu T, Zhan YK, Nie CY, Sun Z. A comparative study on electrosorptive behavior of carbon nanotubes and graphene for capacitive deionization. J Electroanal Chem 2011;653:40–4.
[45] Xu X, Sun Z, Chua DH, Pan L. Novel nitrogen doped graphene sponge with ultrahigh capacitive deionization performance. Sci Rep 2015;5:11225.
[46] Li Q, Jiang R, Dou Y, Wu Z, Huang T, Feng D, Yang J, Yu A, Zhao D. Synthesis of mesoporous carbon spheres with a hierarchical pore structure for the electrochemical double-layer capacitor. Carbon 2011;49:1248–57.
[47] Sun G, Wang J, Li K, Li Y, Xie L. Polystyrene-based carbon spheres as electrode for electrochemical capacitors. Electrochim Acta 2012;59:424–8.
[48] Liu J, Conry TE, Song X, Yang L, Doeff MM, Richardson TJ. Spherical nanoporous LiCoPO4/C composites as high performance cathode materials for rechargeable lithium-ion batteries. J Mater Chem 2011;21:9984–7.
[49] Zhu G-N, Liu H-J, Zhuang J-H, Wang C-X, Wang Y-G, Xia Y-Y. Carbon-coated nano-sized Li4Ti5O12 nanoporous micro-sphere as anode material for high-rate lithium-ion batteries. Energy Environ Sci 2011;4:4016–22.
[50] Liu Y, Chen TQ, Lu T, Sun Z, Chua DH, Pan LK. Nitrogen-doped porous carbon spheres for highly efficient capacitive deionization. Electrochim Acta 2015;158:403–9.

Section C

Capacitive Energy Extraction From Double Layer Expansion (CDLE)

Chapter 5

Capacitive Energy Extraction From Double Layer Expansion (CDLE). Fundamentals of the Method

Doriano Brogioli and Fabio La Mantia
Energy Storage and Energy Conversion Systems, University of Bremen, Bremen, Germany

5.1 Salinity Gradient Power

The increasing energy demands of industrialized society and the issues related to the use of fossil fuels, in particular climate changes, have raised interest in energy sources with a smaller environmental footprint, in terms of renewability, clean operation, decreased impact on ecosystems, and lower use of land. Since it is clear that any energy production process has an environmental footprint and that it is unlikely that a single renewable energy source can cover all the demands, the current trend is to identify an optimal "energy mix" as a trade-off between economic, social, and environmental issues. The footprint of the whole "mix" can be reduced by introducing new, currently untapped renewable resources, each of them optimal for a given geographical location, and covering only an optimal fraction of the total energy need.

Generally speaking, a minimal-footprint energy production process should intercept a naturally occurring process that involves a free-energy loss; the tapping should take place without significantly changing the natural results of the process. In some cases, the free energy dissipation of a natural process is apparent, for example, in the case of a waterfall that can be exploited by a hydropower plant. A less evident case is represented by the estuary of a river, where an entropy production takes place because the salt becomes able to move in a larger volume of water as the river flows into the sea. The entropy increase corresponds to a free energy dissipation of approximately 2.4 kJ/L of fresh water dispersed into the sea [1]. This amount corresponds to the energy dissipated by the same liter of water falling from a height of 240 m, but this

Interface Science and Technology, Vol. 24. https://doi.org/10.1016/B978-0-12-811370-7.00005-X

huge dissipation is not apparent when we observe the calm flow of water in a large estuary of a river.

It is actually possible to intercept the free energy dissipation and tap energy, for example, in the form of electrical work, by interposing a suitable device between the salt and the freshwater, so that the mixing takes place in a controlled way [2, 3]. Various techniques have been developed starting from the oil crisis of the 1970s [4], when alternative energy sources were pursued for economical reasons. The technology is often called salinity gradient power (SGP).

One of the methods proposed for SGP is the so-called pressure-retarded osmosis [5–7], shown in Fig. 5.1. Two compartments, separated by a semipermeable membrane, are filled with salt water and fresh water. Osmosis generates a hydrostatic head of approximately 24 atm in the saltwater compartment and a water flow, which is converted into electrical current by using a turbine. The mixing of solutions is controlled because only water is allowed to move between the solutions of different salt concentrations. The substitution of the salt water, which becomes less concentrated, by seawater does not require significant work and is accomplished by means of the so-called pressure exchanger. At variance with other applications of osmosis, in this technique the osmotic flow is hindered by pressure, hence the name pressure-retarded given to the technique.

An alternative technique is called reverse electrodialysis (RED) [8, 9], shown in Fig. 5.2. Two different kinds of ion-exchange membranes, permeable to positive and negative ions, are alternated and stacked, forming separated compartments. The saltwater and freshwater solutions flow in adjacent compartments. A Donnan potential develops across each membrane, approximately 80 mV, and the various potentials add up along the stack. In the final compartments

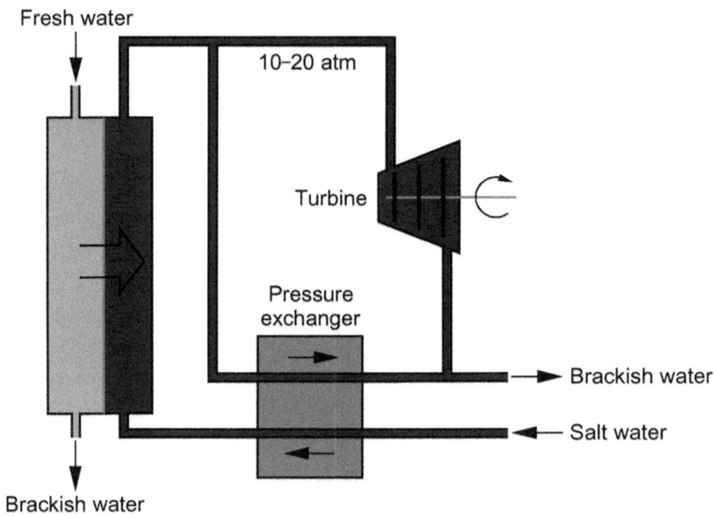

FIG. 5.1 Scheme of the pressure-retarded osmosis technique.

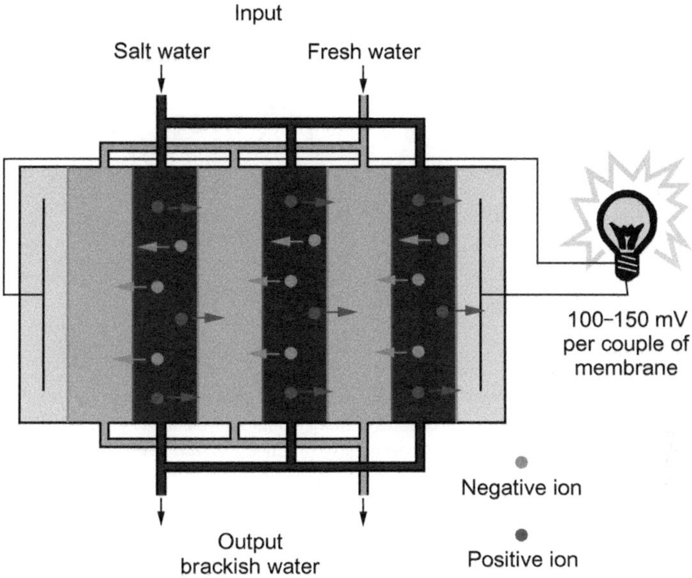

FIG. 5.2 Scheme of the reverse electrodialysis technique.

of the stack, two electrodes dipped in anolyte and catholyte solutions are used to pick up the current. In this technique, the control of the mixing is due to the selectivity of the membranes, which allows the migration of only one species of ions between the solutions of different concentrations.

The SGP technology is at the level of small pilot plants, for both pressure-retarded osmosis [10] and reverse electrodialysis [8, 9]. The main obstacles for the industrial development of these techniques are high cost of the membranes and the short life due to fouling (accumulation of powders and growth of living organisms). The pretreatments, like filtering, are energy-expensive: they are decided on as trade-offs between the necessity of increasing the life of the membranes (thus decreasing the cost for substitution) and the necessity of decreasing the energy cost. Improving the permeability, conductivity, and selectivity of the membranes is a target of fundamental research [8, 9].

Alternative methods have been proposed, without reaching a high technological readiness level. Among them is the so-called vapor pressure method [11], based on the vapor pressure difference between fresh water and salt water at the same temperature, which drives a turbine. Other researchers have focused on the effects of salt solutions on the mechanical properties of polymeric materials, either the variation of length of a collagen fiber [12] or the volume of a hydrogel [13], which lead to production of work upon concentration change. Up to now, none of the mentioned methods has led to industrially relevant developments.

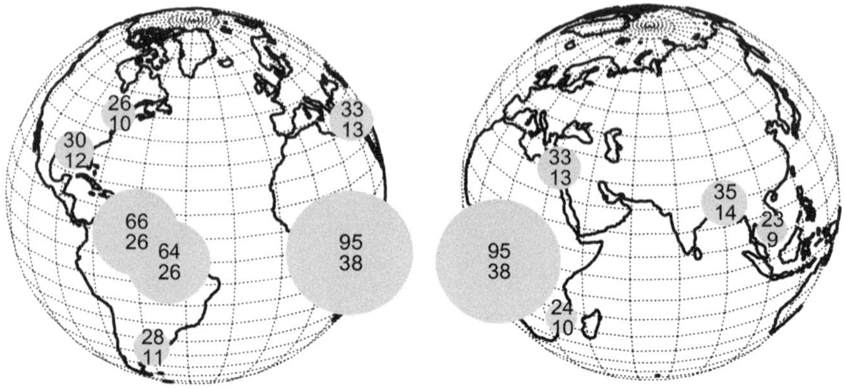

FIG. 5.3 Potential of various rivers of the globe. The *numbers* represent GW; the *upper figure* is the theoretical potential, that is, the available free energy flux, and the *lower figure* is the technical potential, that is, the power that can be economically produced. *(Data from Kuleszo J, Kroeze C, Post J, Fekete BM. The potential of blue energy for reducing emissions of CO_2 and non-CO_2 greenhouse gases. J Integr Environ Sci 2010;7:89–96.)*

The total free energy dissipated by all the rivers of the world is around 2 TW, a relevant part of the global energy need [9, 14]. Fig. 5.3 shows the potential power production from using the fresh water of the main rivers of the world: we can notice that the figures are much larger than the typical power of a nuclear power plant. The power production process would be clean, since the final state is the same as the natural process: that is, the fresh water is eventually mixed with the seawater. However, a large-scale exploitation of the river mouths could result in the devastation of the delicate ecosystems present there; it could be thus better to plan small plants, distributed along the coasts, exploiting smaller amounts of fresh water. The overall potential of the fresh waters is shown in Table 5.1. We can notice that the available power per meter of coast could be significant, for example, for small communities. It is worth noting that the use of fresh water for energy production is not in competition with the availability of fresh water for other uses, such as household or agriculture, because the SGP technologies would use slightly brackish waters, widely available on the coasts, that are not suitable for other uses.

The fact that electrical energy is produced by direct methods (without recourse to electromechanical converters) is a clear advantage over other renewable energy sources. Furthermore, the methods are intrinsically scalable, keeping in mind that the power production in these devices is roughly proportional to their volume. So the final design is easily scaled to the power output requirements.

Until the recent advent of capacitive mixing, or CapMix, techniques [14–16], RED was the only electrochemical technique aimed at energy production by ionic solutions mixing. In RED, solutions with two different concentrations flow at the same time through cell compartments separated

TABLE 5.1 Potential of All the Fresh Water Flux

Region	Number of Rivers	Theoretical Potential (GW)	Technical Potential (GW)	Power Density (Technical) (W/m)
Africa	391	311	190	2600
Asia	1243	374	206	500
Europe	779	94	56	250
North America	1878	321	189	370
Oceania	791	308	195	1240
South America	390	316	148	1020
World	5472	1724	983	640

Notes: The power density is the average power available per meter of coast line.
Data from Kuleszo J, Kroeze C, Post J, Fekete BM. The potential of blue energy for reducing emissions of co2 and non-co2 greenhouse gases. J Integr Environ Sci 2010; 7:89–96.

by ion-exchange membranes. Unlike RED, in CapMix, the solutions are pumped alternately, so that the cell is sequentially charged and discharged in the presence of salty and fresh solutions successively. In this way, the ions are extracted from the first solution and stored in the pore double layers. Upon discharging, ions are released into the dilute solution. Hence the CapMix denomination, as the solutions mixing takes place via the capacitive storage of salt in the electrodes. This charge storage in porous electrodes is the working principle of supercapacitors, and the solutions can be in direct contact with the electrodes [15] or these can be coated by ion-exchange membranes [14]. A related principle applies to the technique called mixing entropy battery, using battery-like (instead of inert) electrodes [17].

It is worth mentioning some advantages of CapMix techniques over RED. One of them is the avoidance (in most designs) of ion-exchange membranes, which are expensive and susceptible to fouling. But even the CapMix approaches requiring membranes [14] are advantageous in that they do not require the use of anolyte and catholyte solutions in the extreme compartments of the stacked cells, as electrical work is produced directly from the salty and fresh feeding solutions [14].

The membrane less CapMix technique [15] is based on the so-called capacitive double layer expansion (CDLE) principle; it can be seen as the rever'se of capacitive deionization [18], which aims at desalinating the salt water. The membrane-CapMix [14] is based on the so-called capacitive Donnan potential (CDP) principle; it can be seen as the reverse of membrane capacitive deionization [19]. Also the mixing entropy battery technique has a desalination counterpart, called desalination battery [20]. This chapter is devoted to the CDLE technique, although the CapMix cycle, described in Section 5.3, is the same for the CDP technique, described in Chapter 7, and for the mixing-entropy battery [17], which is beyond the scope of this book.

5.2 What Happens to the EDL When We Decrease the Concentration of the Solution?

From Chapter 1, we know that the potential versus charge curve of the electrical double layer (EDL) depends on the concentration of the solution. In Fig. 5.4, we show the variation of the voltage across the EDL at fixed surface charge density, calculated by using the GCS (Gouy, Chapman, and Stern) theory expressed by Eqs. (3.1), (3.2) that we report here, assuming a value of 0.1 F/m^2 for C_{St}:

$$\varphi_{GCS}(\sigma,c) = 2\frac{RT}{F}\sinh^{-1}\left(\frac{\sigma}{\sqrt{8\epsilon_0\epsilon_r cRT}}\right) + \frac{\sigma}{C_{St}}. \tag{5.1}$$

We can notice that, for a fixed value of surface charge density, the potential across the EDL increases when the concentration of the salt is decreased. A qualitatively similar behavior is also observed in other models.

When we charge a supercapacitor, made of activated carbon electrodes, in a salt solution, at a given voltage, and then we change the solution to a less

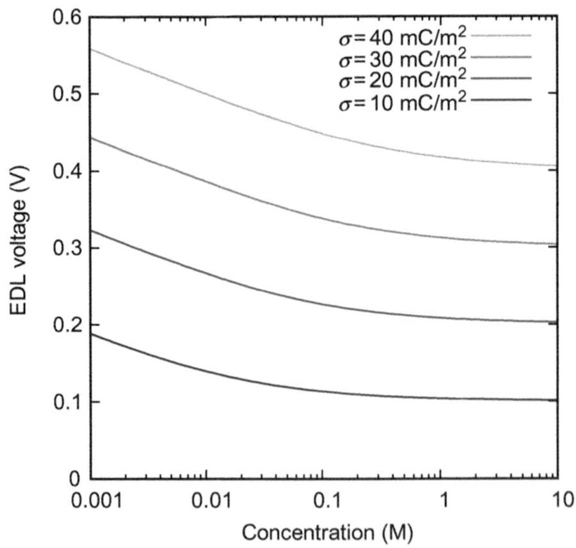

FIG. 5.4 Variation of voltage across the EDL at fixed surface charge density, according to GCS theory.

concentrated one, we actually observe that the potential increases, while the accumulated charge does not change [15]. The experimental results are qualitatively different from the theoretical curves calculated with the models and shown in Fig. 5.4, for reasons that will be highlighted shortly; the variation of the potential versus the salt concentration, at fixed charge, can be found, for example, in Ref. [21].

A qualitative explanation of the phenomenon is given in Fig. 5.5, showing a positively charged conductor dipped into an ionic solution. The conductor attracts a cloud of counterions and repels the co-ions; the picture qualitatively shows this effect, in a more concentrated (left) and diluted (right) solution. When the concentration is decreased (from left to right), the screening effect becomes weaker, that is, the screening length increases: we can thus see that the EDL expands (CDLE). This can be quantitatively seen in the expression of the Debye screening length, Eq. (1.5), Chapter 1. It must be noticed that the figure is only qualitative and is not in scale.

While the dilution takes place, the accumulated charge does not change; thus the electric field between the surface charge of the conductor and the counterion cloud remains approximately constant. However, since the screening length increases, the voltage across the EDL, being the integral of the electric field over the space, also increases.

The increase of voltage upon salinity decrease has a deep analogy with the operation of the electrophorus, a classical device used in past centuries for

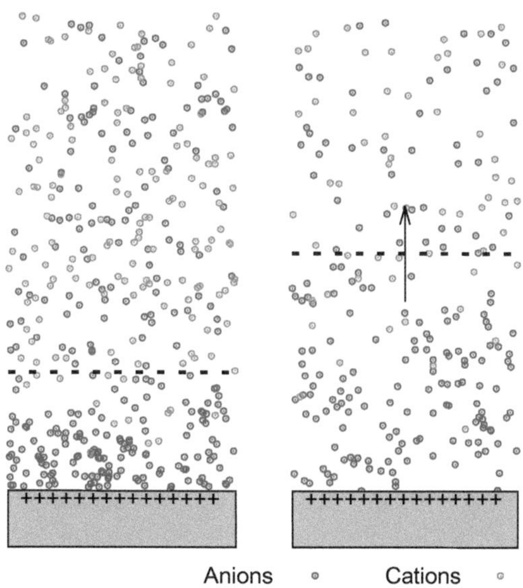

FIG. 5.5 Graphical representation of the effect of the variation of the salt concentration on the EDL. The surface of the conductor is covered by positive charges. The counterion cloud and the repulsion of co-ions can be observed. Passing from the high concentration (*left*) to the low concentration (*right*), the EDL expands.

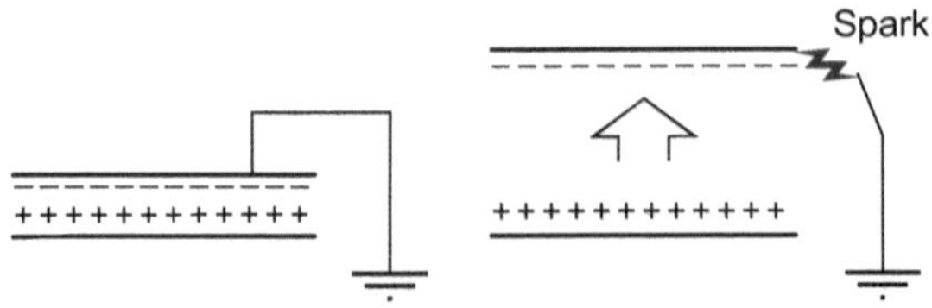

FIG. 5.6 Operation of the electrophorus, an electrostatic device. It is analogous to the capacitive double layer expansion.

experiments on electrostatics. It is shown in Fig. 5.6: a metallic disc is placed on an insulator, charged by friction; its upper surface is shortly grounded, so that a double distribution of opposite charges is formed. Assuming that the potential between the plates is ΔV, the accumulated charge is $Q = C\Delta V$, where C is the capacitance:

$$C = \frac{\epsilon_0 \epsilon_r A}{d}, \tag{5.2}$$

where A is the surface of the plates and d is the spacing. Then the plate is raised. The increase of spacing between the plates decreases C, and thus the voltage increases: $\Delta V = Q/C \propto d$. A spark can thus be obtained between the plate and a grounded object (traditionally, a knuckle of the experimenter).

The plates of the electrophorus correspond to the two, oppositely charged, layers of the EDL. The CDLE phenomenon corresponds to the displacement of the plates.

As already observed, the voltage increase corresponds to an increased electrostatic energy. This can be seen graphically, since it is proportional to the volume in which the electrostatic field is present: as the plates of the capacitor displace from one another, the volume occupied by the electric field increases and consequently also the electrostatic energy.

In the case of the electrophorus, a mechanical work is needed to displace the plates, which attract each other; the work is converted into electrostatic energy. In the case of the EDL, the displacement between the charge distributions is driven by diffusion. It is worth noting that the charges move against the electric force.

By charging a couple of supercapacitor electrodes in seawater, and then substituting the solution with fresh water from a river, we observe a voltage rise on the order of 50–100 mV. This voltage rise corresponds to an increase of electrical energy. In the following section we will see how it is possible to extract this surplus energy by means of a cycle.

5.3 CapMix Cycle

In this section we describe the CapMix cycle, for the production of energy from salinity differences. The cycle is applied to activated carbon electrodes working with the CDLE principle; however, the same cycle is also used for the CDP technique, described in Chapter 7, and for the mixing-entropy battery [17] that is not described in this book.

The basic CapMix method is performed in a cell composed of two porous electrodes placed in contact with the feed solutions, alternately concentrated and diluted. A current is made to flow through the cell in one direction when charging the electrodes, and flows in the opposite direction while discharging them. The cycle is represented in Fig. 5.7. It starts with the cell filled with the high-salinity solution. From this point the steps are as follows:

(A) Electrodes are charged by flowing a current through the solution.
(B) The solution is exchanged with a low-salt one in open circuit.
(C) The circuit is closed and a current flows in the opposite direction to that in step A. The electrodes are discharged.
(D) The salty solution is pumped in the cell, in open circuit.

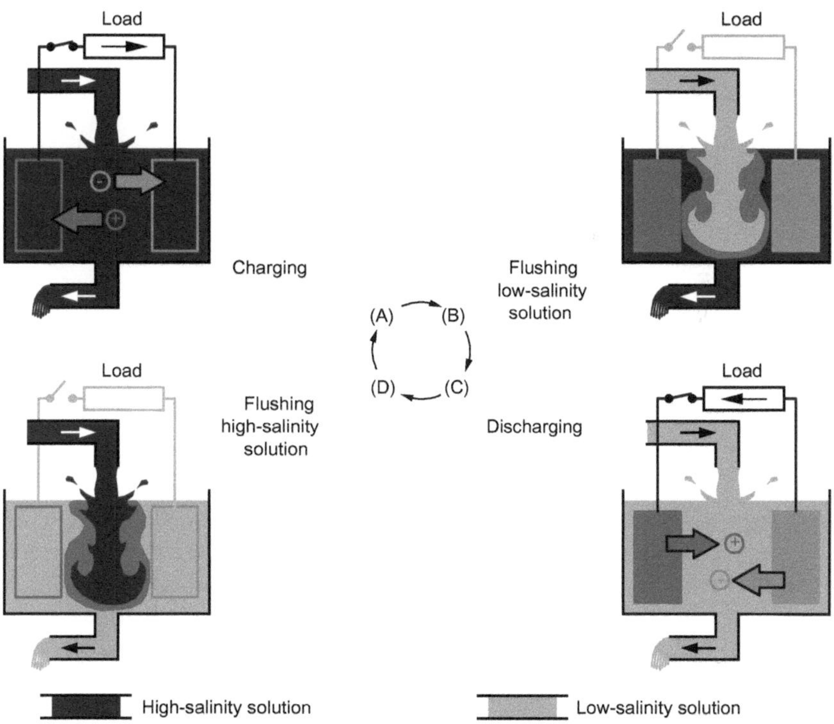

FIG. 5.7 Schematic representation of the four steps of the CapMix cycle. The cycle begins with the cell filled with the high-salinity solution. Step A: A current flows in one direction. Step B: The circuit is opened. The solution in the cell is exchanged with the low-salinity feed solution. Step C: A current flows in the opposite direction with respect to step A. Step D: The circuit is opened. The liquid in the cell is exchanged with the high-salinity feed solution. *(Adapted with the permission of Elsevier from Cipollina A, Micale G, editors. Sustainable energy from salinity gradients. Amsterdam: Woodhead Publishing; 2016, Fig. 6.1.)*

The key aspect of the CapMix cycle is the voltage rise in step B, meaning that the charge is extracted from the cell at a higher voltage than it was stored in it. This CDLE effect is in fact the source of energy production in the cycle, in the form of electrical work in the external circuit. The cycle bears resemblance to heat engines. Thus salt capture will correspond to heating, entrance of dilute salt solution is the equivalent of expansion, the salt release is similar to cooling, and the flow of concentrated salt solution resembles compression.

The time evolution of voltage and current along the cycle is sketched in Fig. 5.8A. Notice the voltage rise during charging and its drop upon discharging. Additionally, the voltage increases while fresh water is pumped in step B and decreases when it is salty water that gets in. The former elevation is typically designated as voltage rise ΔV, and it happens to be an essential quantity of the cycle. In Fig. 5.8B, the voltage is represented as a function of the charge stored in the cell (the integral of the current with time). The surface enclosed in the cycle, evaluated counterclockwise (thanks to the voltage rise, the cycle proceeds counterclockwise in fact, and in this way it delivers work to the external circuit), measures the electrical work W performed by the cell along a cycle:

$$W = \oint_C V(q)\mathrm{d}q. \tag{5.3}$$

We can notice that, thanks to the voltage rise, the cycle is actually performed counterclockwise, thus delivering electrical work to the external circuit.

An approximate evaluation of the work per cycle is given by:

$$W = \Delta V Q, \tag{5.4}$$

where Q is the charge exchanged per cycle. Note that the overvoltages (limit of vanishing current) are neglected here.

5.4 Theory of Salt Adsorption

Charging an EDL involves the formation of a counterion cloud and the repulsion of co-ions from the conductive surface. In principle, the complete screening of the charge on the conductor could take place with a perfect equality of attracted counterions and repelled co-ions, so that the overall EDL is charge-neutral and the average number of ions (positive and negative) near the electrode is equal to the concentration in the bulk, that is, the number of attracted counterions equals the number of repelled counterions. This is actually the prediction of the GCS theory of the EDL in the limit of small charge density, that is, voltage across the EDL is much less than kT/e (see Chapter 1).

At larger voltages, the density of the ions near the electrode becomes larger than the concentration in the bulk, that is, the attracted counterions are more than the repelled co-ions; this is predicted by theory (e.g., GCS)

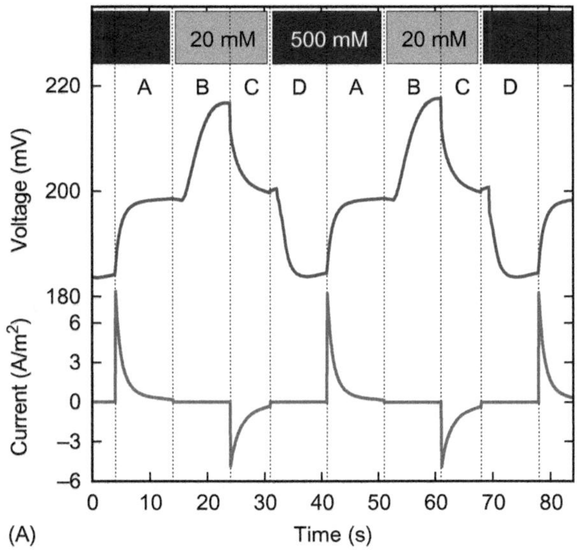

(A)

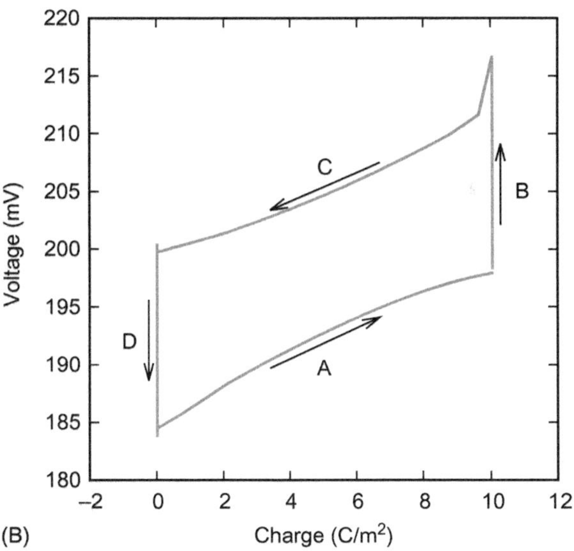

(B)

FIG. 5.8 Cycles in CapMix technique, reported in the voltage versus time graph (A) and voltage versus charge graph (B). The two electrodes are made of activated carbon functionalized with p-toluenesulfonic acid (p-TSA) and polyethyleneimine branched with ethylene diamine (PEI-EN); see Section 5.6.3. The solutions are 20 and 500 mM of NaCl. The current is 2 A/m^2 and the cycle duration is 600 s, with step durations of 120–60–120–60 s. *(Adapted with the permission of Elsevier from Marino M, Misuri L, Jiménez ML, Ahualli S, Kozynchenko O, Tennison S, et al. Modification of the surface of activated carbon electrodes for capacitive mixing energy extraction from salinity differences. J Colloid Interface Sci 2014;436:146–53, Fig. 1.)*

and observed in experiments. When a couple of electrodes are charged in a solution, part of the salt is actually captured and stored in the EDLs. This form of salt adsorption is exploited by the so-called capacitive deionization [18], aiming at removing salt from solutions. In this section we will discuss this phenomenon, which will appear relevant for the CapMix process.

First of all, from thermodynamics we derive a general relation connecting the salt adsorption to the variation of the cell voltage on concentration. This will allow us to calculate the salt adsorption from the potential versus charge relation described in Chapter 2. This relation has been discussed in Ref. [22]; the derivation we present here follows the line of Ref. [23].

Interfacial thermodynamics is now applied to the electrode storing charge in its EDLs. Because faradaic reactions are not considered, ions cannot go from the solution, β, to the solid wall, α, nor can electrons react with species in solution. Excess surface concentrations (mole/m^2) of species i on the solution side, Γ_i^β, and of electrons on the electrode, Γ_e, are defined by (see Fig. 5.9):

$$\Gamma_i^\beta = \int_0^{+\infty} \left(c_i^\beta - c_{i,b}^\beta\right) \mathrm{d}x, \tag{5.5}$$

$$\Gamma_e = \int_{-\infty}^{0} \left(c_e^\alpha - c_{e,b}^\alpha\right) \mathrm{d}x, \tag{5.6}$$

where c is the concentration of the species i or the electrons, and the subscript b indicates the value in the bulk. The Gibbs-Duhem equation for the electrode/electrolyte interface at constant temperature and pressure can be described by [24]:

$$\mathrm{d}\gamma + \Gamma_e \mathrm{d}\mu_e^\alpha + \sum_i \Gamma_i^\beta \mathrm{d}\mu_i^\beta - F\Gamma_e \mathrm{d}\left(\Phi^\alpha - \Phi^\beta\right) = 0, \tag{5.7}$$

where γ is the surface tension, μ_i the chemical potential of species i, Φ^τ the electric potential in the bulk of phase τ, and F the Faraday constant.

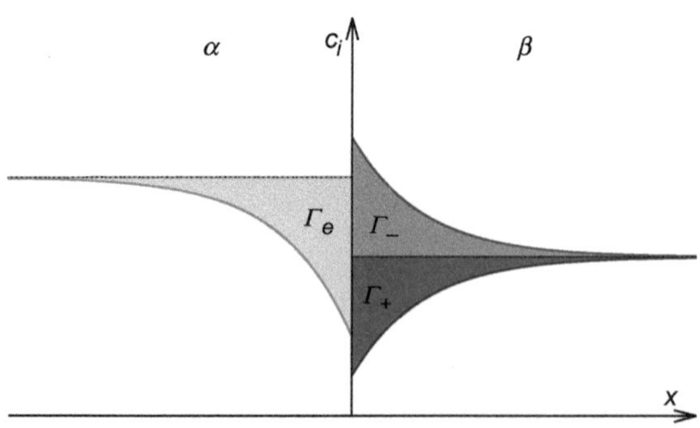

FIG. 5.9 Excess concentrations for electrons in the metal (phase α) and ions in the solution (phase β). The concentration, c_i, tends to change near the interface due to electrostatic and chemical attraction of the species to the surface.

Now we explicitly express the differential form, Eq. (5.7), by assuming that the physical variables are the salt concentration c and the surface charge density on the surface of the electrode, σ, equal to $-F\Gamma_e$:

$$\frac{\partial}{\partial\sigma}\gamma+\sigma\frac{\partial}{\partial\sigma}\left(\Phi^{\alpha}-\Phi^{\beta}\right)=0 \tag{5.8}$$

$$\frac{\partial}{\partial c}\gamma+\sum_i\Gamma_i^{\beta}\frac{\partial}{\partial c}\mu_i^{\beta}+\sigma\frac{\partial}{\partial c}\left(\Phi^{\alpha}-\Phi^{\beta}\right)=0. \tag{5.9}$$

Here we assumed that the chemical potentials of the ions, μ_i^{β}, do not depend on the charge density σ and the chemical potential of the electrons, μ_e^{α}, does not depend either on the charge density σ or on the concentration c. In order to get rid of the unknown surface tension γ, we take the derivative of the first equation with respect to c and the derivative of the second equation with respect to σ, and we subtract them:

$$\sum_i\frac{\partial}{\partial\sigma}\Gamma_i^{\beta}\frac{\partial}{\partial c}\mu_i^{\beta}+\frac{\partial}{\partial c}\left(\Phi^{\alpha}-\Phi^{\beta}\right)=0. \tag{5.10}$$

Notice that this last equation contains the chemical potential of single ions, which is not a physically accessible quantity. This happens because the last equation refers to the potential rise of a single electrode. However, the real quantity that we measure is the voltage of a whole electrochemical cell, measured between electrodes. Thus we write the equations for the two electrodes:

$$\sum_i\frac{\partial}{\partial\sigma'}\Gamma'^{\beta}_i\frac{\partial}{\partial c}\mu_i^{\beta}+\frac{\partial}{\partial c}\left(\Phi^{\alpha'}-\Phi^{\beta}\right)=0, \tag{5.11}$$

$$\sum_i\frac{\partial}{\partial\sigma''}\Gamma''^{\beta}_i\frac{\partial}{\partial c}\mu_i^{\beta}+\frac{\partial}{\partial c}\left(\Phi^{\alpha''}-\Phi^{\beta}\right)=0, \tag{5.12}$$

where the $'$ and $''$ refer to the two different electrodes. We call $\sigma=\sigma'$; since a current changes σ' and σ'' in opposite directions, $\partial/\partial\sigma''=-\partial/\partial\sigma$ (although in general σ' can be different from $-\sigma''$). By subtracting the two equations, we get the following equation for the electrochemical cell consisting of two capacitive electrodes:

$$\sum_i\frac{\partial}{\partial\sigma}\left(\Gamma''^{\beta}_i+\Gamma'^{\beta}_i\right)\frac{\partial}{\partial c}\mu_i^{\beta}+\frac{\partial}{\partial c}\left(\Phi^{\alpha''}-\Phi^{\alpha'}\right)=0. \tag{5.13}$$

From this equation, we see that, to obtain a high value of voltage rise, it is necessary that $\partial\Gamma'^{\beta}_{+}/\partial\sigma'$ and $\partial\Gamma''^{\beta}_{+}/\partial\sigma''$ are as far apart as possible, that is, one should be equal to 1 and the other to 0. From a physical point of view, this requirement is equivalent to saying that one electrode will accumulate only cations and the other only anions. To reach this objective, it is necessary that the two capacitive electrodes are far from the point of zero charge, one above and the other below. Therefore, SGP devices based on symmetric capacitive electrodes need to be precharged, either by means of an external power supply or chemical methods, as will be explained in Section 5.6.3.

In order to relate the adsorption of the ions to the adsorption of the whole salt, first of all focus on a single salt that dissociates into two ions; each mole of salt, dissociated into n_+ and n_- moles of cations and anions, charges z_+ and z_-, respectively. The electroneutrality of the salt solution implies that $n_+z_+ + n_-z_- = 0$. We define the captured salt as Γ_s so that it corresponds to $n_+\Gamma_s$ molar density of cations and $n_-\Gamma_s$ molar density of anions adsorbed in the EDLs.

Before completing the calculation, it is necessary to emphasize that the two quantities $n_+\Gamma_s$ and $n_-\Gamma_s$ do not give the total amount of adsorbed cations and anions. The two double layers are also electroneutral:

$$\sigma' = -F\left(z_+\Gamma'^{\beta}_{+} + z_-\Gamma'^{\beta}_{-}\right), \tag{5.14}$$

$$\sigma'' = -F\left(z_+\Gamma''^{\beta}_{+} + z_-\Gamma''^{\beta}_{-}\right). \tag{5.15}$$

We know that $\sigma' + \sigma''$ does not necessarily vanish; we assume that it is negative (the opposite assumption would lead to the same final results). By summing the last two equations, we see that the total charges in the EDLs do not add up to 0 but to a positive value, hence they cannot come only from the dissociation of the salt. Thus we conclude that there is an amount Γ^{β}_{δ} of cations adsorbed into the EDLs, which is not charge-balanced inside the solution:

$$\sigma' + \sigma'' = -Fz_+\Gamma^{\beta}_{\delta}. \tag{5.16}$$

The unbalanced adsorbed cations Γ^{β}_{δ} can be related to their physical origin. After fabrication, the electrodes are electroneutral. When the electrodes are put into the solution, their surface gets charged because of the dissociation of surface groups. The part of the charge that remains on the surface, possibly different for the two electrodes, leads to the nonvanishing value of $\sigma' + \sigma''$; the dissociated ions are exchanged with the ions in solution and represent the unbalanced adsorbed cations Γ^{β}_{δ}. The unbalanced adsorbed cations Γ^{β}_{δ} can also appear due to the parasitic reactions leading to the self-discharge.

We thus conclude that the total excess molar densities of anions and cations are:

$$\Gamma'^{\beta}_{+} + \Gamma''^{\beta}_{+} = n_+\Gamma^{\beta}_{s} + \Gamma_{\delta}, \tag{5.17}$$

$$\Gamma'^{\beta}_{-} + \Gamma''^{\beta}_{-} = n_-\Gamma^{\beta}_{s}. \tag{5.18}$$

These equations are compatible with Eq. (5.15), when the proper definition of Γ_{δ}, that is, Eq. (5.16), is taken into account.

As already underlined, when σ' changes by an amount $\delta\sigma$, then σ'' changes by an opposite amount. Thus the derivative of the left-hand side of Eq. (5.16) with respect to σ vanishes. We thus conclude that $\partial\Gamma_\delta/\partial\sigma$ also vanishes.

Together with this consideration, by substituting the values of Eq. (5.18) into Eq. (5.13) we get:

$$\frac{\partial}{\partial\sigma}\Gamma_s^\beta\frac{\partial}{\partial c}\mu_s^\beta+\frac{\partial}{\partial c}\left(\Phi^{\alpha''}-\Phi^{\alpha'}\right)=0, \tag{5.19}$$

where the overall chemical potential of the salt is defined as $\mu_s^\beta = n_+\mu_+^\beta + n_-\mu_-^\beta$.

The term $\partial\Gamma_s^\beta/\partial\sigma$ represents the quantity of salt captured by a given charge; it is called charge efficiency in the literature [25]. The derivative with respect to c is the differential voltage rise δV; it is connected to the voltage rise ΔV by the relation:

$$\delta V(c)=\frac{\mathrm{d}}{\mathrm{d}c}\Delta V(c',c,\mu)\bigg|_{c'=c}, \tag{5.20}$$

where the voltage rise ΔV is expressed as a function of the concentrations c' and c of the salt in the concentrated and dilute solutions.

We see that the two properties, that is, capturing of salt and presenting a voltage rise, are connected and always simultaneously present [22].

Eq. (5.19) can be used to calculate the salt adsorption from the voltage versus charge curve. For example, we consider the GCS theory expressed by Eq. (5.1) and we write it for a couple of electrodes:

$$\Delta\varphi_{GCS}(\sigma,c)=4\frac{RT}{F}\sinh^{-1}\left(\frac{\sigma}{\sqrt{8\epsilon_0\epsilon_r cRT}}\right)+2\frac{\sigma}{C_{St}}. \tag{5.21}$$

By using Eq. (5.19), and assuming that the solution is ideal, we calculate the differential salt adsorption $\partial\Gamma_s^\beta/\partial\sigma$:

$$\frac{\partial}{\partial\sigma}\Gamma_s^\beta=\frac{1}{F}f\left(\frac{\sigma}{\sqrt{8\epsilon_0\epsilon_r cRT}}\right), \tag{5.22}$$

where

$$f(x)=\frac{x}{\sqrt{x^2+1}}. \tag{5.23}$$

Since $f(x)\rightarrow 1$ for $x\rightarrow +\infty$, we see that $\partial\Gamma_s^\beta/\partial\sigma\rightarrow 1/F$ for high surface density: this means that the EDL tends to store as many moles of salt as moles of electrons [26].

It is possible to find an approximate integral version of Eq. (5.19), following Ref. [22]. First of all, we assume that the solutions are ideal, and we introduce the differential charge efficiency $\lambda = F\partial\Gamma_s^\beta/\partial\sigma$, that is, the quantity of salt captured by a given change in charge [25, 26]:

$$\frac{\partial}{\partial c}\left(\Phi^{\alpha''} - \Phi^{\alpha'}\right) = -\frac{RT}{F}\frac{2\lambda}{c}. \tag{5.24}$$

The voltage rise in a CapMix process can be obtained by integrating this equation with respect to c, from c_A to c_B, representing, respectively, the concentrations of the fresh and salt water, for example, c_A=20 mM and c_B=500 mM. In most practical cases, the variation of λ in the range (C_A, C_B) is negligible with respect to the variation of $1/c$, and thus we can assume that λ is constant with respect to the integration variable c. The result of the approximate integration is:

$$\Delta V \simeq 2\frac{RT}{F}\ln\left(\frac{c_B}{c_A}\right)\lambda, \tag{5.25}$$

where ΔV is the voltage rise upon the concentration decrease [22]. This expression clarifies the relation between the voltage rise and efficiency in capturing salt, physical parameters that are at the basis of CapMix and capacitive deionization, respectively.

5.5 Thermodynamics of the CapMix Process

The production of energy from the CapMix cycle relies on the occurrence of the voltage rise when the solution concentration is decreased; in Section 5.4 we have seen that the ability of the electrodes to capture the salt always associates with the voltage rise. With reference to the CapMix cycle shown in Fig. 5.7, it is assumed that ions are captured at step A and released at step C. A loss of mixing free energy ΔG occurs due to the transfer of ions from the high-salinity solution to the low-salinity solution. This loss of free energy will appear as electrical work of the external circuit, and drives the ions against the electric field during the potential rise. Hence the process takes place close to reversibility. In order to evaluate ΔG, we consider that the solutions of interest in SGP typically contain a salt dissociated in two monovalent ions. Let V_i be the solution volumes (i equals 1 for the high-salinity solution and 2 for the low-salinity one). The respective initial and final concentrations are c_i^I and c_i^F. It is further assumed that the solutions are ideal and that the volumes do not change during the process. With this, the mixing free energy ΔG is:

$$\Delta G = 2RTV_1c_1^F\log c_1^F - 2RTV_1c_1^I\log c_1^I + 2RTV_2c_2^F\log c_2^F - 2RTV_2c_2^I\log c_2^I, \tag{5.26}$$

where R is the gas constant and T is the temperature. For a small number of transferred salt moles Δn:

$$\Delta G = RT\Delta n \log\left(\frac{c_1^I}{c_2^I}\right), \tag{5.27}$$

which represents the thermodynamically available free energy that can be transformed into electrical work. This is so because, as mentioned, the processes are close to reversible in the limit of the vanishing current. This (quasi)reversibility implies that the CapMix cycles can be "reversed" becoming in this manner a desalination device [18–20].

In the limit of vanishing current, the electrical work can be equated to ΔG, and from Eqs. (5.4), (5.27), we can estimate the voltage rise ΔV.

$$\Delta V = \lambda\frac{2RT}{F}\log\left(\frac{c_1^I}{c_2^I}\right), \tag{5.28}$$

where λ is the desalination charge efficiency [26] described in Section 5.4, and meaning the number of moles captured in the electrodes for one mole of electrons through the electrodes. This approximate relation, which has been obtained under very specific hypotheses, is equal to Eq. (5.25).

The CapMix cycles described so far can be described as Stirling-like, whereby steps A and C occur at nearly constant concentration (similar to the isothermal stage in heat engines), and the solutions exchange in steps B and D brings about a free energy loss, associated with the rapid mixing in the electrodes, favored by their porous structure. A Carnot-like, hence more thermodynamically efficient, situation [27] can be imagined in which a strong desalination/salt release in step B/D is achieved in the absence of liquid flow.

The energy production of the CapMix cycle has been explained up to now in terms of the second principle of thermodynamics: the produced work is no more than the consumed free energy. However, it is also possible to wonder how the conservation of energy is respected:

$$W = \Delta H - \Delta Q. \tag{5.29}$$

In the case of sodium chloride solutions or seawater/river water, the mixing enthalpy ΔH is small with respect to the mixing Gibbs free energy $\Delta G = \Delta H - T\Delta S$: a negligible heat is adsorbed from the environment when two solutions at different concentrations are irreversibly mixed (e.g., by putting them in a beaker and stirring). When the CapMix process takes place isothermally, the produced work is thus close to the adsorbed heat: $W = -\Delta Q$.

This picture can be qualitatively described by saying that the produced work completely comes from heat adsorbed from the environment (first principle):

$$W = -\Delta Q. \tag{5.30}$$

The entropic part of the mixing free energy, $-T\Delta S$, is the coin we have to spend in order to enable the extraction of work from heat (second principle):

$$\frac{\Delta Q}{T} - T\Delta S < 0. \tag{5.31}$$

For sodium chloride solutions, the production of work mainly comes from the mixing entropy. However, for different salts and at high concentration, the mixing enthalpy is possibly relevant and can contribute significantly to the production of work [28–30].

The performed analysis helps us in selecting the best electrodes for the CapMix cell. As in supercapacitors, the cell works by almost reversibly accumulating and releasing charge. However, it is essential that one of the electrodes can absorb the cations (sodium, say) from the feed solution, and the other should do the same with anions (chloride, for example). This is the way activated carbon electrodes work, as they get oppositely polarized during the charging process. In contrast, this is not true in the case of batteries, as exemplified by lithium ion ones: the positive electrode is a material capable of intercalating lithium, and the negative one is graphite (also a lithium-intercalating material). As a whole, no salt capturing takes place, and such a combination would not work in a hypothetical CapMix cycle based on lithium chloride solutions.

5.6 Voltage Rise of the Electrodes

The typical concentration of seawater is around 500–600 mM. River water is much less concentrated, but, at the river mouth, water is often not fresh but slightly brackish, and the water with lower concentration is preferentially used for other applications (e.g., agriculture); a typical concentration in CapMix proof of concept is usually around 20–25 mM. This (relatively high) concentration is chosen not only because it is representative of available river water in some real situation, but also because of a more technical reason: the increase of the resistance due to a lower concentration is not counter balanced by the advantage given by the increase of the energy density and voltage rise. Otherwise noted, high- and low-concentration solutions are 500 and 20 mM of NaCl.

For the CDLE CapMix technique, the voltage rise of the electrodes, when the solution is changed from high-concentration to low-concentration water, has typical values in the range between 20 and 100 mV. In this section we discuss how the voltage rise depends on the charging state of the electrodes and the materials used.

5.6.1 Voltage rise versus base voltage graph

The voltage rise is experimentally evaluated by measuring the difference between the voltage of a cell (or the potential of an electrode) when it is filled with high- and low-concentration solutions. The voltage rise is measured at various charge density σ of the electrodes, changed by charging the electrodes.

It is well known that the supercapacitors undergo self-discharge, and this effect is particularly pronounced in real water solutions containing dissolved oxygen. However, the time scale on which the voltage rise develops is on the order of seconds or minutes, while the time scale of the self-discharge is more than one order of magnitude slower [21]: this allows the evaluation of the voltage rise at different electrode charging.

By monitoring the potential of each electrode of the cell, with respect to a reference electrode, it can be observed that the electrodes tend to a particular spontaneous potential, likely representing the mixed potential of a set of parasitic reactions. At this spontaneous potential the self-discharge, or better the leakage current, vanishes.

The experimental evaluation of the charge density σ is difficult, so the electrodes are brought to a given potential φ_0 (this is typically the potential in the high-concentration solution) by means of a potentiostat, so that a given charge density will be generated. Theoretical expressions for φ can then be used for evaluating the relationship between the voltage rise $\Delta\varphi$ and the base potential φ_0, which can be expressed in parametrical form:

$$\varphi_0 = \varphi\left(\sigma, c_1^l\right), \tag{5.32}$$

$$\Delta\varphi = \varphi\left(\sigma, c_2^l\right) - \varphi\left(\sigma, c_1^l\right), \tag{5.33}$$

σ being the curve parameter. This theoretical expression will be compared with experimental results (in the graph of Figs. 5.10 and 5.12), in particular for the GCS model, Eq. (5.1). We note that the offset along the abscissa (base voltage) is arbitrary, since the potentials are measured with respect to a given reference electrode (Ag/AgCl in the case of the figure). An analogous concept will be applied to the voltage of the complete cell, whereby the cell voltage rise is evaluated experimentally as a function of the cell base voltage.

Whatever the mathematical model of the EDL (e.g., GCS), it is possible to evaluate an upper limit of the electrode potential rise by using Eq. (5.19). The maximum value of the desalination charge efficiency γ is 1; we can thus impose $\partial\Gamma_s^\beta/\partial q < 1/F$:

$$-\frac{\partial}{\partial c}\left(\phi^{\alpha''} - \phi^{\alpha'}\right) < \frac{\partial}{\partial c}\mu_s^\beta. \tag{5.34}$$

By integrating in c:

$$\Delta V < \Delta\mu_s^\beta = 2\frac{RT}{F}\log\frac{c_H}{c_L}, \tag{5.35}$$

where the equality holds under the assumption of the ideality of the solution. The right-hand side is equivalent to the expression of the potential in the Nernst law; this potential will be called thermodynamic potential in the following. Its value is approximately 160 mV for $c_H = 500$ mM and $c_L = 20$ mM.

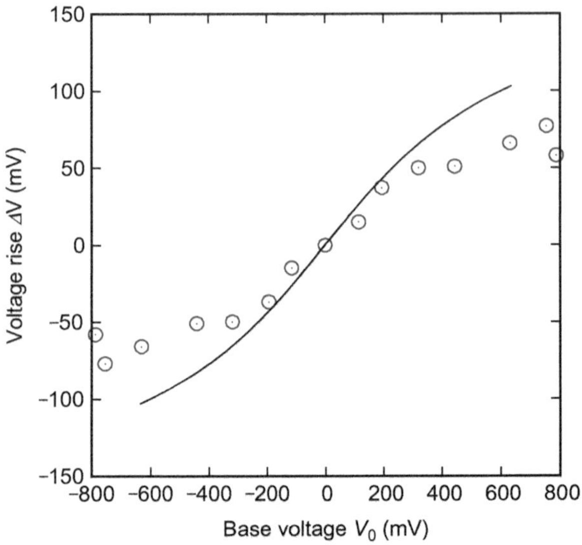

FIG. 5.10 Voltage rise versus base voltage of a cell with equal electrodes made of activated carbon. *(Adapted with the permission of Elsevier from Marino M, Misuri L, Jiménez ML, Ahualli S, Kozynchenko O, Tennison S, et al. Modification of the surface of activated carbon electrodes for capacitive mixing energy extraction from salinity differences. J Colloid Interface Sci 2014;436:146–53, Fig. 4.)*

5.6.2 Simple activated carbon

In the first CapMix implementations with CDLE [15, 31], identical activated carbon electrodes were used. The cell was initially ("externally") charged to a given base voltage <1 V to avoid water dissociation. The measured voltage rise depends on the charge in the EDLs, and hence on the base voltage. The rise was found to be about 10% of the base voltage for NaCl solutions at 25 and 600 mM [15], or 20 and 500 mM [32, 33], and 25%–30% at 1 and 500 mM [31].

A typical dependence of the voltage rise on the base voltage is shown in Fig. 5.10. We see that the voltage rise saturates at a value of approximately 80 mV, which is half the thermodynamic value. In the figure, we also see the curve obtained by the GCS theory. The theoretical curve is in fairly good agreement for small potentials, but saturates at a much higher voltage rise (corresponding to the thermodynamic value).

Experiments performed with the externally charged approach [15, 31–33] produced a power on the order of 10 mW/m^2. The main limitation of this approach is the self-discharge of the cell. This requires a bias current continuously provided to the cell; this operation consumes an amount of energy ΔV δQ, where δQ is the leaked charge, typically of the same order of magnitude as the energy produced through the CapMix cycle itself [21]. The phenomenon of self-discharge (or charge leakage) is well known also in the supercapacitor or capacitive deionization [34] technologies. Its origin lies on the recombination of charges in the EDL, as a consequence of redox reactions, water molecules. It consists of a recombination of the charges across the double layer: it is due to redox reactions associated with surface groups, water molecules, dissolved oxygen, or other gases. The problem cannot be avoided, although, as shown in the following section, there are methods to circumvent it.

5.6.3 Activated carbon with surface modifications

The various materials and electrode assemblies used in CapMix are sketched in Fig. 5.11: (A) simple activated carbon, initially charged by an external power supply; (B) charged adsorbed molecules on the carbon surface; (C) surface of the activated carbon coated by charged polymers; (D) membrane-based CapMix, with ion-exchange membranes (composed of charged molecules) macroscopically covering the electrodes. In the cartoons, both charges and

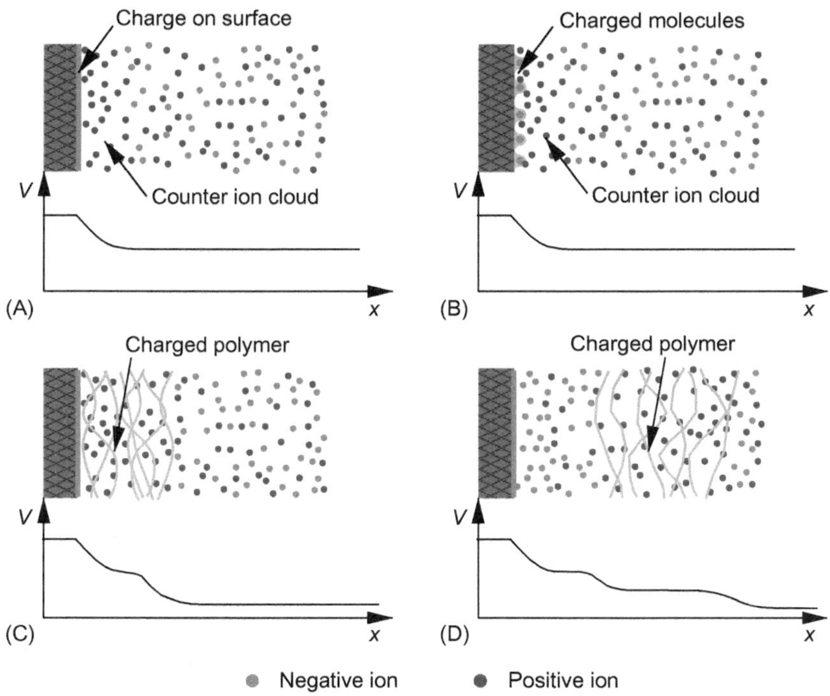

FIG. 5.11 Sketch of the charge distribution in the various electrodes used in the CapMix technique: (A) simple activated carbon, initially charged by an external power supply; (B) charged adsorbed molecules on the carbon surface; (C) coating of the surface by means of charged polymers; (D) membrane-based CapMix, with ion-exchange membranes (composed of charged molecules) macroscopically covering the electrodes. *(Adapted with the permission of Elsevier from Cipollina A, Micale G, editors. Sustainable energy from salinity gradients. Amsterdam: Woodhead publishing; 2016, Fig. 6.5.)*

molecules on the electrodes and the solution are shown, although the detail of the porous structure of the activated carbon is neglected. In these simplified sketches, the charges attracting the counterions are located at increasing distance from the electrode surface, namely: inside the electrode (A), on the surface (B), distributed close to the surface (C), macroscopically separated from the electrode (D).

The adsorbed molecules (see Fig. 5.11B) build a charged layer with density $\sigma_{adsorbed}$, located between the electrode surface (charge density σ) and the counterion cloud [35]. In order to fulfill electroneutrality requirements, the diffuse counterion cloud will have a charge density amounting to $-\sigma - \sigma_{adsorbed}$. In addition, an electric field will be established between the ideal surfaces with σ and $\sigma_{adsorbed}$, so that the potential $\varphi_{adsorbed}$ is now:

$$\varphi_{adsorbed}(\sigma, c_S) = \varphi_{GCS}(\sigma + \sigma_{adsorbed}, c_S) - \alpha \frac{\sigma_{adsorbed}}{C_{St}} \tag{5.36}$$

where $0 \leq \alpha \leq 1$ is a parameter depending on the position of the charged layer [35]. By substituting this equation into Eq. (5.33), it can be seen that the effect of the adsorbed charges is the shift of the potential rise versus base potential curve along the base potential axis. This is experimentally illustrated in Fig. 5.12.

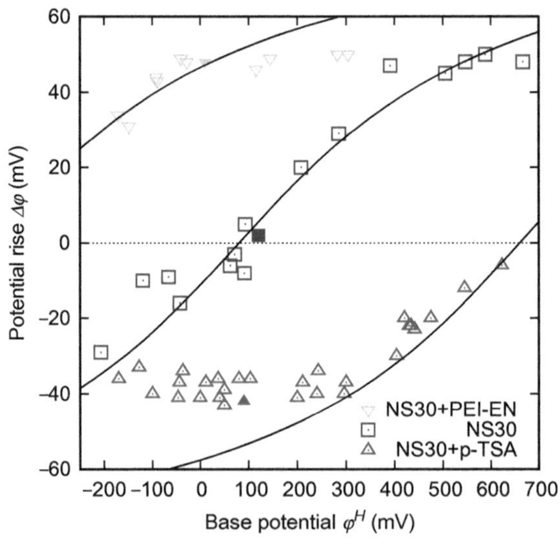

FIG. 5.12 Experimentally measured potential rise $\Delta\varphi$ of NS30 activated carbon electrodes modified by adsorption of charged molecules, p-TSA and PEI-EN, with NaCl solutions at 20 and 500 mM. The *solid symbols* refer to the materials at their spontaneous potentials. *(Adapted with the permission of Elsevier from Marino M, Misuri L, Jiménez ML, Ahualli S, Kozynchenko O, Tennison S, et al. Modification of the surface of activated carbon electrodes for capacitive mixing energy extraction from salinity differences. J Colloid Interface Sci 2014;436:146–53, Fig. 5.)*

Fig. 5.11C is a scheme of the case of CapMix electrodes coated with charged polymers [36]. It is a kind of intermediate situation between a thin film of adsorbed molecules and a membrane layer, as the polyelectrolyte coating is apparently less perm-selective than actual membranes (as used in RED, for instance). It has been demonstrated [36] that the changes in salinity affect both the EDLs on the carbon pores and on the outer boundary of the polymer layer. Nevertheless, also in this case one finds that the limit for the potential rise in highly charged interfaces is the one given by thermodynamic considerations (Eq. 5.35).

Coming back to the problem of self-discharge, it was demonstrated in [21] that it can be mitigated by the observation that the process stops when the electrode potential reaches a given value, known as spontaneous potential. If the electrodes are identical, their spontaneous potentials will coincide, and in the external charging scheme both will reach the same potential rise, giving a negligible value for the cell voltage rise. One clear way to avoid this problem is the use of differently treated electrodes on each side of the cell. For instance, if the carbons used on each of the electrodes have different origins, or if they have been subjected to different activation methods, they will reach different spontaneous potentials, and, most interesting, a cell voltage rise will be developed without external charging [21]. The method proved successful; rendering of power densities on the order of 50 mW/m^2 has been reported [21]. The disadvantage is that the production of such tailor-made carbons is expensive and not suited to industrial upscaling.

The progress associated with the use of couples of different electrodes was already clear in capacitive deionization [37–39]. This line of research has been advanced by a number of works and the procedure of external charging, by means of a power supply, was in fact changed into a chemical charging. Various coatings were devised for that purpose, including: charged molecules (positive on one electrode and negative on the other) [40]; coating with a polyelectrolyte layer [36]; oxidizing the activated carbon for creating negatively charged groups on its surface [41].

Fig. 5.12 shows some results. It contains the voltage rise for different base potentials corresponding to an untreated carbon material (Norit Super30), together with the same material after adsorption of p-toluenesulfonic acid (p-TSA, a negatively charged molecule) or polyethyleneimine branched with ethylene diamine (PEI-EN, a positively charged polymer). Note that the three curves appear similar, except for a horizontal shift, a consequence of the fact that although the spontaneous potentials (solid symbols) are practically identical for the three materials, their voltage rises differ in dependence on the pretreatment. The combination of electrodes with adsorbed p-TSA and PEI-EN molecules yields a total cell voltage rise of approximately 80–90 mV by exchanging 20 and 500 mM NaCl solutions. A similar result was reported for the polymer coating [36] technique; when the pair of electrodes were oxidized versus pristine activated carbon, the voltage rise was 53 mV.

The previously mentioned shift has been explained in terms of EDL properties according to Eq. (5.36). In fact, a comparison between the experimental data of Fig. 5.12 and the GCS and mD models shows a good agreement for low-voltage rises. However, experimental results show saturation at roughly 40–50 mV [40] when exchanging 20 and 500 mM NaCl solutions, in contrast with the 80 mV (the thermodynamic limit, see Eq. 5.35) theoretically predicted. The observed saturation at 40–50 mV is likely to be originated by an incomplete solution exchange, in an out-of-equilibrium situation. In spite of that, it must be pointed out that the chemical charging method yields a stable and reproducible power density (e.g., 30 mW/m^2 with a 100 μm thick film [40]) or approximately 0.2 mW/g.

5.7 Kinetic Aspects of CDLE

The activated carbon electrodes used in supercapacitors and for other electrochemical applications, such as CapMix, have a porous structure. Porous materials have a large surface for the formation of EDLs, resulting in a huge capacitance. However, the porous structure has the drawback that the mass transport of ions along the pores is hindered. The resulting resistance in ion transport makes the distribution of current inhomogeneous along the thickness of the electrodes [42]. A very simplified model representing some features of a porous material is the transmission line model of de Levie [43–45]: the pore is decomposed into several segments, each representing the capacitance C of a part of the EDLs; the current flows through distributed resistances, ρ_E and ρ_S representing the conductor and the ionic solution, respectively. This equivalent circuit expresses the different delays at which the various parts of the pore are charged.

A more complete approach allows evaluation of the effect of salt capture and release along the thickness of the electrodes [46, 47]. This approach is schematically shown in Fig. 5.13: the electrode is represented as a continuum

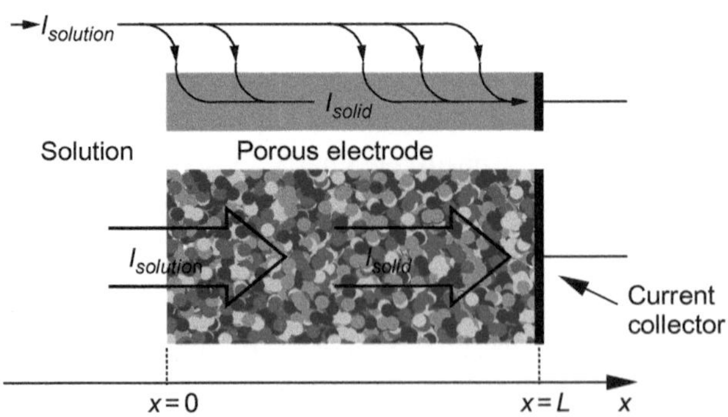

FIG. 5.13 Schematic representation of the model of the current in the porous material [46]. Along the thickness of the porous electrode, the current passes from the solution to the solid.

medium; the currents flowing in the solution $I_{solution}$ and in the solid I_{solid} are parameterized as a function of the coordinate x. The mass balance equation in the porous electrode is given by:

$$\frac{\partial c}{\partial t} = D\frac{\partial^2 c}{\partial x^2} - \frac{1}{a}\cdot\frac{\partial \Gamma}{\partial t}, \tag{5.37}$$

where c is the concentration of the solution, D is the diffusion coefficient, a is a parameter describing the geometrical properties of the material, and Γ is salt adsorption at the interface. The charge balance equation is:

$$\frac{\partial I_{solution}}{\partial x} = -\frac{1}{a}\frac{\partial q}{\partial t}, \tag{5.38}$$

where q is the charge stored in the EDL. The potential and the current are connected by Ohm's law:

$$I_{solution} = -2\lambda c \frac{\partial \varphi_{solution}}{\partial t}, \tag{5.39}$$

where λ is the equivalent ionic conductivity and $\varphi_{solution}$ is the potential on the solution. Finally, q, Γ, and $\varphi_{solution}$ are connected through the EDL model (e.g., GCS).

Boundary equations must be added to Eqs. (5.37), (5.38), together with the equations for describing the stagnant diffusion layer in the solution near the electrode. Numerical solutions for the case of CapMix cycles have been calculated, with the aim of optimizing the CapMix cycles [48, 49]. Here we report a summary of some of the findings.

Fig. 5.14 (from Ref. [49]) shows the evolution of the potential and of the concentration at $x = 0$, that is, at the macroscopic interface between the solution and the porous material, during the CapMix cycle. The vertical branches represent the constant-concentration steps A and C, in which the surface charge q is changed from 0.2 to 0.25 (a dimensionalized quantities [49]). We see that the results calculated for steps B and D do not follow the voltage versus concentration curve expected in equilibrium conditions (curves $|q| = 0.20$ and $|q| = 0.25$, dashed magenta lines) but rather approach the curves for intermediate values of charge density (namely $|q| = 0.24$ and $|q| = 0.22$, in this case green solid lines), for reaching eventually the equilibrium value.

This apparent discrepancy in the charge has been explained [49] in terms of charge redistribution. When the solution is changed, the part of the electrode at $x = 0$ is first affected by the concentration change. Due to the CDLE, the voltage across the EDL is no longer uniform (higher where the concentration is lower), leading to a current in the pores and in the solid that redistributes the charge (although there is no net current flowing in the external circuit). For this reason, the charge at the surface during the solution changes appears different from the average charge at equilibrium.

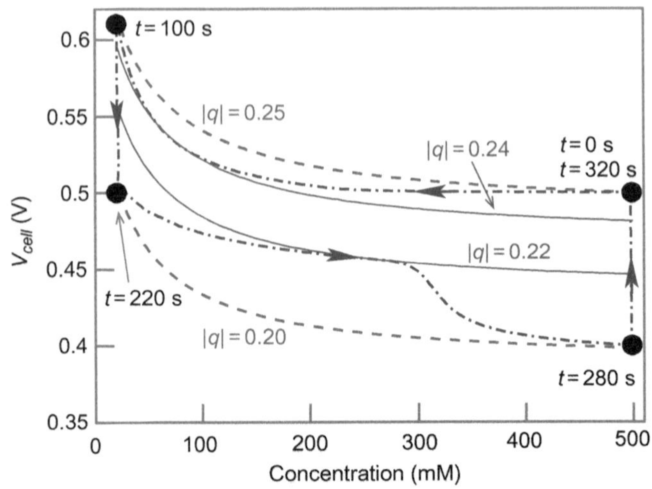

FIG. 5.14 Evolution of the potential and of the concentration at the macroscopic interface between the solution and the porous material, during the CapMix cycle. *(Adapted with the permission of Elsevier from Rica RA, Ziano R, Salerno D, Mantegazza F, Bazant MZ, Brogioli D. Electro-diffusion of ions in porous electrodes for capacitive extraction of renewable energy from salinity differences. Electrochim Acta 2013;92:304–14, Fig. 5.)*

An equivalent effect takes place during charge and discharge steps, A and C [48]: the actual concentration in the electrode is affected by salt adsorption and release, which takes place nonuniformly along the thickness of the electrode because the current is mostly concentrated on the surface. This phenomenon affects the overvoltage during the CapMix cycle and must be considered when an optimization of the cycle is attempted.

An effect that is clearly observed in CapMix experiments is the different speed at which the voltage changes after changing the concentration of the solution: when the solution concentration is increased (step D), the change of the potential is much faster than in step B. This effect is obtained in numerical calculations [48], also when the redistribution of charge along the electrode thickness is made negligible. The calculations show that the strong asymmetry is due to the highly nonlinear dependence of the voltage across the EDL on the concentration [48].

5.8 Comparison with Other SGP Techniques

The performances of various realizations of the CapMix technique are reported in Table 5.2 and compared with the two more traditional techniques, pressure-retarded osmosis and reverse electrodialysis. Besides the CDLE technique, described in this chapter, the data for CapMix include the CDP technique and the technique with polymer-coated electrodes, which is intermediate between CDP and CDLE.

TABLE 5.2 Comparison Between Various Realizations of the CapMix Technique and Other SGP Techniques

Technique	Voltage Rise (mV)	Power Density (mW/m^2)	Reference
CapMix			
CDLE, simple activated carbon electrodes	33	7	[15]
CDLE, electrodes with adsorbed molecules	90	30	[40]
Electrodes with polymer coating	90	12	[36]
Battery-based	140	105	[17]
CDP (membrane-based)	140	200	[50]
Other techniques			
Reverse electrodialysis	(160 mV)	1300	[51]
Pressure-retarded osmosis	(24 atm)	2300	[51]

Notes: The voltage rise is evaluated between NaCl solutions at concentrations of 20 and 500 mM. For the two traditional techniques, pressure-retarded osmosis and reverse electrodialysis, we report the theoretical pressure and cell voltage, respectively.

In terms of operation, we can notice that CapMix is cyclical while RED is continuous. Thus CapMix has an additional difficulty, connected with the need to periodically switch the solution inside the cell, and a loss of free energy due to the unwanted mixing of the solutions in the dead volumes.

The main advantage of CDLE CapMix with respect to the membrane-based techniques (CDP CapMix, RED, and pressure-retarded osmosis) is that it avoids the use of the expensive perm-selective or semipermeable membranes. With respect to the battery-like electrodes [17], the activated carbons (used in CDLE CapMix) have faster kinetics [52]; this corresponds to the well-known fact that batteries have a larger capacitance but a longer charging and discharging time with respect to supercapacitors.

In Fig. 5.4, it can be observed that, when the concentration of the solution is increased at fixed charge density, the potential of the EDL reaches a plateau. This means that the voltage rise of the CapMix CDLE technique is expected to be lower for more concentrated solutions. It is worth noting that a similar difficulty is present also in RED, due to the loss of perm-selectivity at high concentration; instead, the use of battery-like electrodes in the context of a CapMix cycle [17] should be effective also at high concentrations.

Although presently the performance of RED is superior to that of CapMix regarding power density and efficiency, it can be expected that this is compensated for in large-scale implementations. In that case, the materials needed for CapMix might be available at a much reduced price, making it worthwhile to strengthen research efforts in capacitive mixing techniques.

References

[1] Semiat R. Energy issues in desalination processes. Environ Sci Technol 2008;42(22): 8193–201. https://doi.org/10.1021/es801330u.

[2] Pattle RE. Production of electric power by mixing fresh and salt water in the hydroelectric pile. Nature 1954;174:660.

[3] Logan BE, Elimelech M. Membrane-based processes for sustainable power generation using water. Nature 2012;488(7411):313–9.

[4] Norman RW. Water salination: a source of energy. Science 1974;186:350–3.

[5] Levenspiel O, de Nevers N. The osmotic pump. Science 1974;183:157–60. https://doi.org/10.1126/science.183.4121.157.

[6] Loeb S. Osmotic power plant. Science 1975;189:654–5. https://doi.org/10.1126/science.189.4203.654.

[7] Chung TS, Li X, Ong RC, Wang QGH, Han G. Emerging forward osmosis (FO) technologies and challenges ahead for clean water and clean energy applications. Curr Opin Chem Eng 2012;1(3):246–57. https://doi.org/10.1016/j.coche.2012.07.004.

[8] Weinstein JN, Leitz FB. Electric power from differences in salinity: the dyalitic battery. Science 1976;191:557–9.

[9] Post JW, Hamelers HVM, Buisman CJN. Energy recovery from controlled mixing salt and fresh water with a reverse electrodialysis system. Environ Sci Technol 2008;42:5785–90.

[10] Gerstandt K, Peinemann KV, Skilhagen SE, Thorsen T, Holt T. Membrane processes in energy supply for an osmotic power plant. Desalination 2008;224(1–3):64–70.

[11] Olsson M, Wick GL, Isaacs JD. Salinity gradient power: utilizing vapour pressure differences. Science 1979;206:452–4.

[12] Sussman MV, Katchalsky A. Mechanochemical turbine: a new power cycle. Science 1970;167:45–7. https://doi.org/10.1126/science.167.3914.45.

[13] Zhu X, Yang W, Hatzell MC, Logan BE. Energy recovery from solutions with different salinities based on swelling and shrinking of hydrogels. Environ Sci Technol 2014;48(12): 7157–63. https://doi.org/10.1021/es500909q.

[14] Sales BB, Saakes M, Post JW, Buisman CJ, Biesheuvel PM, Hamelers HV. Direct power production from a water salinity difference in a membrane-modified supercapacitor flow cell. Environ Sci Technol 2010;44:5661. https://doi.org/10.1021/es100852a.

[15] Brogioli D. Extracting renewable energy from a salinity difference using a capacitor. Phys Rev Lett 2009;103:058501. https://doi.org/10.1103/PhysRevLett.103.058501.

[16] Rica RA, Ziano R, Salerno D, Mantegazza F, van Roij R, Brogioli D. Capacitive mixing for harvesting the free energy of solutions at different concentrations. Entropy 2013;15(4): 1388–407. https://doi.org/10.3390/e15041388.

[17] La Mantia F, Pasta M, Deshazer HD, Logan BE, Cui Y. Batteries for efficient energy extraction from a water salinity difference. Nano Lett 2011;11:1810–3. https://doi.org/10.1021/nl200500s.

[18] Oren Y. Capacitive deionization (CDI) for desalination and water treatment—past, present and future (a review). Desalination 2008;228:10–29. https://doi.org/10.1016/j.desal.2007.08.005.

[19] Biesheuvel PM, van der Wal A. Membrane capacitive deionization. J Membr Sci 2010;346(2):256–62. https://doi.org/10.1016/j.memsci.2009.09.043.

[20] Pasta M, Wessells CD, Cui Y, La Mantia. A desalination battery. Nano Lett 2012; 12:839–43. https://doi.org/10.1021/nl203889e.

[21] Brogioli D, et al. Exploiting the spontaneous potential of the electrodes used in capacitive mixing technique for extraction of salinity-difference energy. Energy Environ Sci 2012;5(12):9870–80. https://doi.org/10.1039/C2EE23036D.

[22] Rica RA, Ziano R, Salerno D, Mantegazza F, Brogioli D. Thermodynamic relation between voltage-concentration dependence and salt adsorption in electrochemical cells. Phys Rev Lett 2012;109:156103. https://doi.org/10.1103/PhysRevLett.109.156103.

[23] Cipollina A, Micale G. Sustainable energy from salinity gradients. Amsterdam: Woodhead Publishing; 2016.

[24] Bard AJ, Faulkner LR. Electrochemical methods. Fundamentals and applications. New York: John Wiley and Sons, Inc.; 2001.

[25] Biesheuvel PM, van Limpt B, van der Wal A. Dynamic adsorption/desorption process model for capacitive deionization. J Phys Chem C 2009;113:5636–40.

[26] Zhao R, Biesheuvel PM, Miedema H, Bruning H, van der Wal A. Charge efficiency: a functional tool to probe the double layer structure inside porous electrodes, and application in the modeling of capacitive deionization. J Phys Chem Lett 2010;1:205–10. https://doi.org/10.1021/jz900154h.

[27] Boon N, van Roij R. 'Blue energy' from ion adsorption and electrode charging in sea and river water. Mol Phys 2011;109(7–10):1229–41. https://doi.org/10.1080/00268976.2011.554334.

[28] Marino M, Misuri L, Carati A, Brogioli D. Boosting the voltage of a salinity-gradient-power electrochemical cell by means of complex-forming solutions. Appl Phys Lett 2014;105: 033901. https://doi.org/10.1063/1.4890976.

[29] Marino M, Misuri L, Carati A, Brogioli D. Proof-of-concept of a zinc-silver battery for the extraction of energy from a concentration difference. Energies 2014;7(6):3664–83. https://doi.org/10.3390/en7063664.

[30] Carati A, Marino M, Brogioli D. Thermodynamic study of a distiller-electrochemical cell system for energy production from low temperature heat sources. Energy 2015;93(1): 984–93. https://doi.org/10.1016/j.energy.2015.09.108.

[31] Brogioli D, Zhao R, Biesheuvel PM. A prototype cell for extracting energy from a water salinity difference by means of double layer expansion in nanoporous carbon electrodes. Energy Environ Sci 2011;4:772–7. https://doi.org/10.1039/C0EE00524J.

[32] Jimenez ML, Fernandez MM, Ahualli S, Iglesias G, Delgado AV. Predictions of the maximum energy extracted from salinity exchange inside porous electrodes. J Colloid Interface Sci 2013;402:340–9. https://doi.org/10.1016/j.jcis.2013.03.068.

[33] Iglesias GR, Fernández MM, Ahualli S, Jiménez ML, Kozynchenko OP, Delgado AV. Materials selection for optimum energy production by double layer expansion methods. J Power Sources 2014;261:371–7. https://doi.org/10.1016/j.jpowsour.2013.12.125.

[34] Porada S, Zhao R, van der Wal A, Presser V, Biesheuvel PM. Review on the science and technology of water desalination by capacitive deionization. Prog Mater Sci 2013;58(8): 1388–442. https://doi.org/10.1016/j.pmatsci.2013.03.005.

[35] Brogioli D, Ziano R, Rica RA, Salerno D, Mantegazza F. Capacitive mixing for the extraction of energy from salinity differences: survey of experimental results and electrochemical models. J Colloid Interface Sci 2013;407:457–66. https://doi.org/10.1016/j.jcis.2013.06.050.
[36] Ahualli S, Jiménez ML, Fernández MM, Iglesias G, Brogioli D, Delgado AV. Polyelectrolyte-coated carbons used in the generation of blue energy from salinity differences. Phys Chem Chem Phys 2014;16:25241–6. https://doi.org/10.1039/c4cp03527e.
[37] Avraham E, Noked M, Cohen I, Soffer A, Aurbach D. The dependence of the desalination performance in capacitive deionization processes on the electrodes PZC. J Electrochem Soc 2011;158(12):168–73. https://doi.org/10.1149/2.078112jes.
[38] Gao X, Omosebi A, Landon J, Liu K. Enhancement of charge efficiency for a capacitive deionization cell using carbon xerogel with modified potential of zero charge. Electrochem Commun 2014;39:22–5. https://doi.org/10.1016/j.elecom.2013.12.004.
[39] Gao X, Omosebi A, Landon J, Liu K. Surface charge enhanced carbon electrodes for stable and efficient capacitive deionization using inverted adsorption-desorption behavior. Energy Environ Sci 2015;8:897–909. https://doi.org/10.1039/C4EE03172E.
[40] Marino M, Misuri L, Jiménez ML, Ahualli S, Kozynchenko O, Tennison S, et al. Modification of the surface of activated carbon electrodes for capacitive mixing energy extraction from salinity differences. J Colloid Interface Sci 2014;436:146–53. https://doi.org/10.1016/j.jcis.2014.08.070.
[41] Hatzell MC, Raju M, Watson VJ, Stack AG, van Duin ACT, Logan BE. Effect of strong acid functional groups on electrode rise potential in capacitive mixing by double layer expansion. Environ Sci Technol 2014;48(23):14041–8. https://doi.org/10.1021/es5043782.
[42] Newman J, Thomas-Alyea KE. Electrochemical systems. Hoboken, NJ: John Wiley and Sons, Inc.; 2004.
[43] de Levie R. On porous electrodes in electrolyte solutions: I. Capacitance effects. Electrochim Acta 1963;8:751–80.
[44] de Levie R. On porous electrodes in electrolyte solutions—IV. Electrochim Acta 1964;9:1231–45. https://doi.org/10.1016/0013-4686(64)85015-5.
[45] de Levie R. The influence of surface roughness of solid electrodes on electrochemical measurements. Electrochim Acta 1965;10:113–30. https://doi.org/10.1016/0013-4686(65)87012-8.
[46] Newman J, Tobias CW. Theoretical analysis of current distribution in porous electrodes. J Electrochem Soc 1962;109:1183–91. https://doi.org/10.1149/1.2425269.
[47] Biesheuvel PM, Bazant MZ. Nonlinear dynamics of capacitive charging and desalination by porous electrodes. Phys Rev E 2010;81:031502. https://doi.org/10.1103/PhysRevE.81.031502.
[48] Rica RA, Brogioli D, Ziano R, Salerno D, Mantegazza F. Ions transport and adsorption mechanisms in porous electrodes during capacitive-mixing double layer expansion (CDLE). J Phys Chem C 2012;116(32):16934–8. https://doi.org/10.1021/jp3059849.
[49] Rica RA, Ziano R, Salerno D, Mantegazza F, Bazant MZ, Brogioli D. Electro-diffusion of ions in porous electrodes for capacitive extraction of renewable energy from salinity differences. Electrochim Acta 2013;92:304–14. https://doi.org/10.1016/j.electacta.2013.01.063.
[50] Liu F, Schaetzle O, Sales BB, Saakes M, Buisman CJN, Hamelers HVM. Effect of additional charging and current density on the performance of capacitive energy extraction based on Donnan potential. Energy Environ Sci 2012;5:8642–50. https://doi.org/10.1039/C2EE21548A.

[51] Ramon GZ, Feinberg BJ, Hoek EMV. Membrane-based production of salinity-gradient power. Energy Environ Sci 2011;4:4423–34. https://doi.org/10.1039/C1EE01913A.

[52] Marino M, Misuri L, Ruffo R, Brogioli D. Electrode kinetics in the "capacitive mixing" and "battery mixing" techniques for energy production from salinity differences. Electrochim Acta 2015;176:1065–73. https://doi.org/10.1016/j.electacta.2015.07.069.

Further Reading

[53] Kuleszo J, Kroeze C, Post J, Fekete BM. The potential of blue energy for reducing emissions of CO_2 and non-CO_2 greenhouse gases. J Integr Environ Sci 2010;7:89–96.

Chapter 6

Capacitive Energy Extraction From CDLE: Implementation

Guillermo R. Iglesias* and María M. Fernández†

*Department of Applied Physics, University of Granada, Granada, Spain

†Department of Fluidic and Energy Science, Central America University "José Simeón Cañas", San Salvador, El Salvador

6.1 Introduction

As reviewed before in this book, theories of electrical double layer (EDL) capacitance [1, 2] for the capacitive double layer expansion (CDLE) method [3] predict that the voltage rise when salty and fresh water solutions are exchanged is about 80 mV on each electrode. Simultaneously, the surface charge density is expected to change by 0.05 C/m^2. As a result, the energy produced in one cycle will be ideally around 4 mJ/m^2 of EDL; if a 1000 m^2/g carbon is used as electrode material, we could get something like 4 J/cycle, using just 1 g carbon. With a 10% efficiency and a cycle duration of 100 s, the power available will not be negligible at all, reaching 2 W/m^2. Contrary to RED and PRO [4, 5], such predicted amounts of energy and power for CDLE are much larger than those experimentally achievable. In this situation, it is clear that *implementation research* should play a crucial role in order to increase the amount of energy and power available [6].

In this chapter, we will consider how CDLE can be implemented in practice. We will start by a description of the materials and methods that are most often used. Afterward, it will be necessary to investigate some *operational issues* of the different cycle stages, going from the effect of the potential difference applied during the charging step to the optimum initial charging time [7]. Moreover, a deep analysis of the charge leakage is required in order to understand its effect and maintain acceptable energy extraction cycles.

On the other hand, we will also consider the change in capacitance associated with another parameter apart from salinity variations, namely *temperature variations*: the capacitance decreases when temperature is raised, and hence a cycle is possible in which some charge is put on the electrode at a certain temperature and returned at a higher one. It has been experimentally

Interface Science and Technology, Vol. 24. https://doi.org/10.1016/B978-0-12-811370-7.00006-1

demonstrated that it is possible to obtain energy from electric double layers if these are successively contacted with solutions at different temperatures. In addition, temperature and salinity variations can be conveniently combined to maximize the electrode potential increase and energy production [8].

It can be expected that the characteristics, in particular hydrophilicity, of the pore walls and pore structure, will be determinant. This is inspired by research on the electrochemical response of supercapacitors, which has been found to be strongly dependent on the degree of matching of the pore and solvated ion sizes [9]. Additionally, *different salinity difference sources* have been analyzed giving way to brine and brackish water solutions that could be used to maximize the extracted power and reduce its environmental impact at the same time.

Finally, we will also tackle a different but fundamental aspect toward CDLE real implementation: exploring new methods for *stacking individual capacitive mixing (Capmix) cells* in order to find the best configuration of a multielectrode setup for maximizing the energy and power outputs [10].

6.2 The Capmix Cell

A typical Capmix cell will serve as a compartment where water can be exchanged between two facing activated carbon electrodes [11]. Fig. 6.1A shows a picture of the one used in our laboratory. A graphite disk 2 cm in diameter is used as the electrical contact (Fig. 6.1B). On this, an activated carbon film of the same area is placed (Fig. 6.1C). The two electrodes are separated by a 600-μm plastic film spacer.

Activated carbon is most suited for fabrication of EDL capacitors. From an electrochemical point of view, this element is relatively unreactive and thus has a potential voltage range of almost ideal polarizability, approaching 1.0 V in aqueous solution [12–14]. Moreover, activated carbon films are

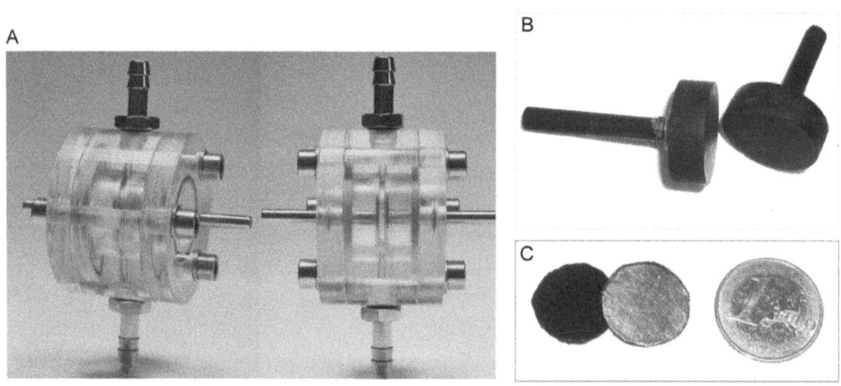

FIG. 6.1 (A) Picture of the capmix cell. Components of the capmix cell: (B) graphite collector and (C) carbon films.

TABLE 6.1 Specific Surface Area of Activated Carbon Samples

Carbon Samples	Specific Surface Area (m^2/g)
SC-1	1423
SC-2	1489
SR-03	1870
SR-23	927
SR-51	959

chosen in order to maximize charge exchange and the huge interfacial area of nanoporous carbon particles fulfills this condition. The carbon films of Fig. 6.1C were made as described in [15].

Numerous activated carbon samples have been tested for CDLE optimum implementation. The ones used in our laboratory were supplied by Mast Carbon Ltd (UK). In the following list, we sum up the description of these mentioned carbon samples and Table 6.1 shows their specific surface area:

- *Samples SC-1 and SC-2*: SC-1 was prepared by activation (40% burn-off) of mesoporous carbon beads (250–500 μm in diameter) derived from mesoporous phenolic resin beads prepared by the patented method of Tennison et al. [16]. The carbon sample SC-2 was prepared by oxidation of SC-1 in air at 320°C to give additional 15% burn-off and decorate the surface with oxygen-containing functional groups [16].
- *Samples SR-03, SR-23, and SR-51*: SR-03 was prepared by activation at 900°C (50% burn-off) of microporous granular carbonisate of phenolic resin [16]. Activation of the same carbonisate with carbon dioxide at 800°C in the presence of catalytic quantities of calcium oxide resulted in preferential development of small mesopores to give samples SR-23 (35% burn-off) and SR-51 (44% burn-off).

The capmix concept relies on the EDL structure, in which only Coulombic interactions and diffusion are considered. However, in real systems other water structure-mediated interactions can take place, such as hydrogen bonding, hydrophilic-hydrophobic interaction, etc., this modifying the relation between charge and potential and affecting the capacitive mixing cell cycle [17]. Hence an exhaustive characterization of the charge at the Stern layer as well as a chemical analysis of the ionic species and complexes present in the river and seawater is a fundamental step in material characterization. For these reasons, in addition to the usual electrophoretic mobility determinations, the wettability of the carbon samples should be characterized, typically via contact angle determinations for water droplets deposited on the carbon film.

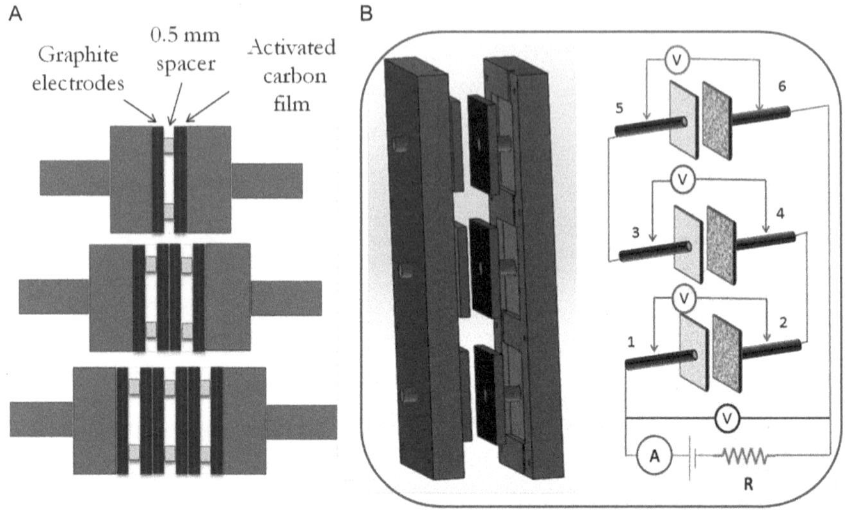

FIG. 6.2 (A) Sketch of the association of individual cells used in stack 1. (B) (*Left*) Sketch of stack 2 configuration. (*Right*) Electrical connection in series of the sealed stack. *(Reproduced with permission from Iglesias GR, Ahualli S, Fernández MM, Jiménez ML, Delgado ÁV. Stacking of capacitive cells for electrical energy production by salinity exchange. J Power Sources 2016;318:283–90. Copyright 2016 Elsevier Ltd.)*

As we have introduced, capmix stacking is needed to explore methods for adding individual cells in order to increase the amount of energy that can be extracted from blue energy by capmix methods. Fig. 6.2 shows two different stack designs that have been tested. Fig. 6.2A shows a scheme of the individual cell association designated as stack 1, based on the device shown in Fig. 6.1.

With the aim of analyzing the behavior of every unit in the stack, we designed the stack 2 (Fig. 6.2B). This stack is basically the same configuration as stack 1, but individual determinations of voltage are possible for each cell. Adjacent cells are externally connected, and all the electrodes are in contact with the same solution bath, but now adjacent cells are externally connected.

6.3 Suggestions for Methodology

Two kinds of capmix experiments can be typically carried out in order to perform energy extraction cycles. In the first case, a power source is employed to fix the potential and, in the second one, a current source guarantees that the closed circuit processes are performed at constant current. Fig. 6.3 shows an example of a CDLE experimental cycle where the two mentioned modes are carried out: constant voltage and constant current. In both cases, it can be seen how the capmix cell of Fig. 6.1A is connected to an external circuit where the voltage between both electrodes and the current going in or out of the cell is recorded as a function of time.

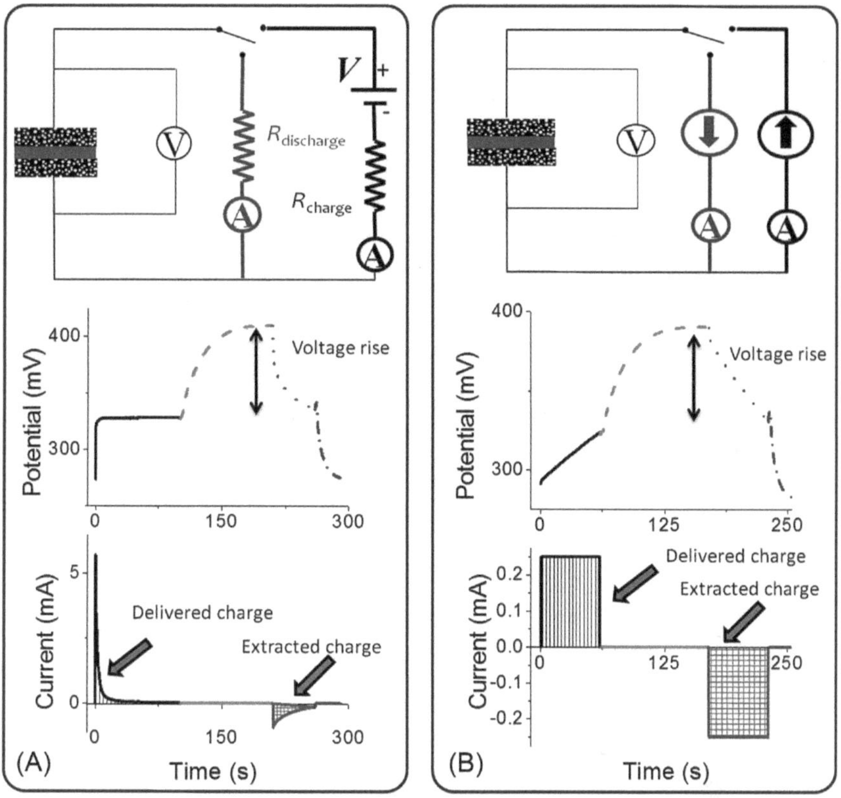

FIG. 6.3 External capmix circuit and experimental measurements during a CDLE cycle: current intensity versus time and potential difference between the cell electrodes versus time for constant voltage (A) and constant current (B) mode. Sample: SR51 film.

- *Constant voltage mode (CV).* In step 1, electrodes immersed in salty water are connected to a voltage source set at a certain potential V. This voltage will be referred to as the working potential. Ions redistribute in the solution until a certain electronic charge grows at the electrode surface and the potential of each electrode reaches half of the source potential value $\Psi = V/2$. Second, the circuit is opened and salty water is replaced by fresh water, giving rise to double layer expansion. As shown in the graph, the electrode surface potential increases ($\Delta\Psi$) because the capacitance decreases at a constant surface charge. Afterward, the circuit is closed again. The cell, which is now at a higher potential, is discharged through a resistance, bringing about a decrease of both capmix cell voltage and charge. Finally, the circuit is opened and the fresh water is replaced by salty water again, leading to a further decrease of the voltage below the source value. Fig. 6.3A shows CV mode external circuit and at the same time, a representation of the potential and current as a function of time.

- *Constant current mode (CC).* This mode is represented in Fig. 6.3B. A constant current is first applied to the capmix cell, driving the potential to a certain value. The circuit is opened and fresh water provokes a potential rise in the cell. Finally, the circuit is closed again and the current is forced in the opposite direction until the potential is again close to the same initial value. The end step involves the entrance of salt water in the cell and the subsequent decrease of the potential is recorded.

It is worth noting that in the discharging step (dotted lines in Fig. 6.3) there is a sudden decrease of the potential difference, followed by a slower decay. The first corresponds to the voltage drop inside the solution, which depends on the value of the solution resistance and the discharge resistor (or equivalently, discharge current value). These ohmic losses can be minimized to a certain limit by choosing an appropriate discharge resistance or current and by decreasing the distance between electrodes. Hence part of the energy is not extracted but dissipated in the solution. The slower stage of voltage decrease corresponds to the capacitive discharging of the cell through the external and internal resistances.

At any time the charge can be calculated by integration of the current versus time representations in Fig. 6.3. Indeed, the black-shadowed area corresponds to the delivered charge by the voltage or current source during the initial charging step, and the green-shadowed one indicates the extracted charge during the discharging process after the voltage rise in fresh water.

The energy as a function of time can be calculated as:

$$E(t) = -\int_0^t V(t')I(t')dt' \tag{6.1}$$

where V and I are the measured voltage and current and hence $E(t)$ is the actual energy that exits the cell. Finally, the power is obtained as the average slope in the curve $E(t)$ in successive cycles.

6.4 Optimum Working Conditions

Although one could think that the potential rise is an intrinsic property of the EDL, the effect of the potential difference applied during the charging step might also affect the whole process even if the voltage is kept below the minimum value for faradaic reactions. In fact, such an effect was observed in the earliest implementations of the method [18]. Below that value we have a whole range of working potentials where current leakage is always present but in different ways.

Current leakage consists of the tendency of the voltage difference between the charged electrodes to move toward a given "spontaneous" potential. Fig. 6.4 shows this effect on three consecutive cycles at constant voltage (A) and constant current (B) modes, both representing the cell potential versus the

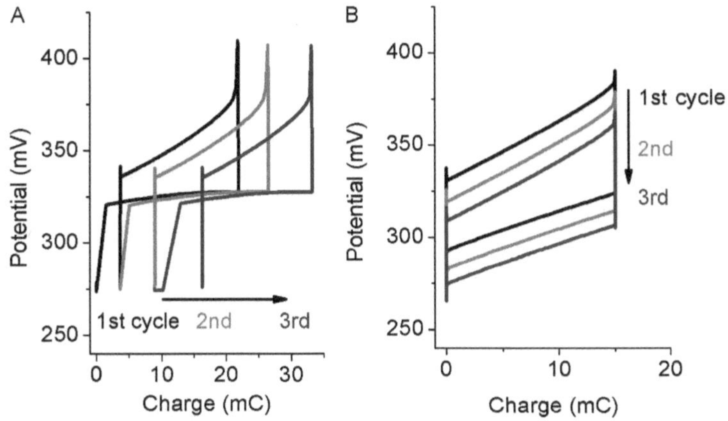

FIG. 6.4 CDLE cycles represented by cell potential versus charge at constant voltage (A) and at constant current (B) mode. Sample: SR51 film.

measured charge. Although caused by the same effect, opposite behaviors can be found on the two representations: in constant voltage mode, cycles move toward the right, to higher charge values, while at constant current, they shift toward lower potentials as is indicated by the respective arrows.

As said, it should be noted that these behaviors are caused by the same unavoidable current leakage effect. When the voltage is fixed (Fig. 6.4A), the charge that is stored in the cell (delivered charge) is not completely recovered in the discharging step (extracted charge) and these unbalanced values produce a cycle shift of approximately 0.01 mC/s. On the other hand, when the charge is fixed in constant current cycles (Fig. 6.4B), the leakage current results in a decrease of the cell potential of about 0.05 mV/s. This effect manifestly points out the current leakage effect, which causes an important reduction in terms of the total extracted energy per cycle.

The selection of the optimum working potential has important effects on the CDLE performance. In fact, an appropriate selection is able to reduce the mentioned charge leakage and optimize the extracted power. This is illustrated in Fig. 6.5, where CDLE cycles are plotted in the form of cell voltage versus charge (A), together with the voltage rise, transferred and delivered charge (B), for different working potentials [6]. This figure demonstrates that:

- The voltage rise always increases for the studied working potentials, that is, less than 550 mV.
- Delivered and extracted charges also increase from around 2 mC to approximately 15 mC for the highest potential. This fact produces wider cycles, as can be seen in Fig. 6.5A.

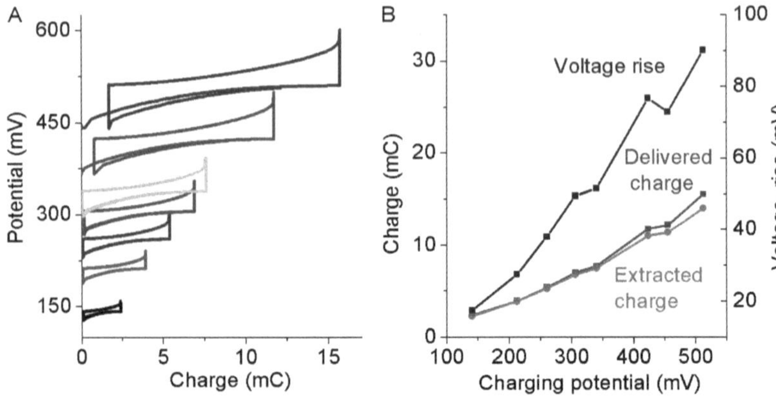

FIG. 6.5 (A) Measured charge as a function of cell voltage of the CDLE cycle for different working voltages in cells built with SR51 carbon films. CDLE cycles represented by cell voltage as a function of measured charge (B). Voltage rise, extracted, and delivered charge obtained for different charging source potential. Sample: SR51 film.

TABLE 6.2 Extracted Power for Different Working Potentials

Working Potential (mV)	Extracted Power (mW/m^2)
140	0.04
212	0.32
260	0.42
306	0.74
340	1.04
423	0.57
455	0.18
512	–

- The difference between the extracted and delivered charge undergoes a similar dependence on working potential, but even increases with it. This situation, far from being favorable, is the main cause of the CDLE cycle leakage. The reason is that we are far from the ideal situation of recovering the whole delivered charge.

These results offer a clue regarding working conditions in the CDLE technique. When the potential used for charging is raised, the self-discharge rate also increases [7], affecting the extracted power, as can be seen in Table 6.2 where at the highest potential (512 mV) there is no positive extracted power. On the contrary, the optimum working potential is found at an intermediate potential (340 mV) where the potential rise as well as the extracted and delivered charge are not as high as they could be but are able to compensate for the mentioned leakage.

It has also been found that initial charging is equally determinant for the CDLE cycle performance and should be considered carefully. The charge redistribution [19–21] effect takes place at the porous electrodes during the initial charging process: shortly after disconnecting for solution exchange, the cell potential undergoes a rapid decrease, prior to a more monotonous trend. This is the charge redistribution inside the pores, as in fact demonstrated by the absence of such rapid decline if the charging times are enlarged [13]. Therefore, a protocol was followed to charge the capmix cell. This is resumed in the following stages [6]:

- Charging cycles should include repeated cycles of flowing salt water at open circuit during 30 seconds and charging at constant voltage for 500 seconds.
- It has been found that charging overnight improves the results by decreasing current leakage.
- After an appropriate charging process the voltage decay rate in open circuit is below approximately 0.5 mV/s.

6.5 Temperature Influence on CDLE

In a number of industrial processes, water is used as coolant and mixed afterward in the cold water reservoir. The temperature gradient between them can be quite high in the case of thermal power plants, although temperature differences between the water input and output are hardly above 20°C, after being refrigerated in the power plant. A different, natural case is that of waters near geysers or in volcanic areas, which can reach 85°C or more.

In this section we propose to show that it is possible to take advantage of solution temperature differences in the direct production of electrical energy, as a consequence of the capacitance changes induced in the EDLs of microporous electrodes by a temperature difference between the water streams in successive stages of the cycle. In fact, a change in the capacitance of the EDL is produced not only by changing the salinity of the solution in contact with the interface but, additionally, by changing the permittivity of the solution by, for instance, increasing or reducing its temperature. Thus, an increase in temperature produces a decrease of the electric permittivity of water, and thus a decrease of the differential capacitance of the EDL, and a subsequently larger electrode potential for given charge.

In the case where there are no salinity differences between the solutions used, but just temperature variations, it was proposed in [8] to denominate DLPE (or double layer permittivity exchange) to this technique. In Fig. 6.6, we present a cycle designed for obtaining electric energy from just temperature differences. At the first stage the electrodes are externally connected to a battery with potential difference V in the presence of cold water. At equilibrium, the

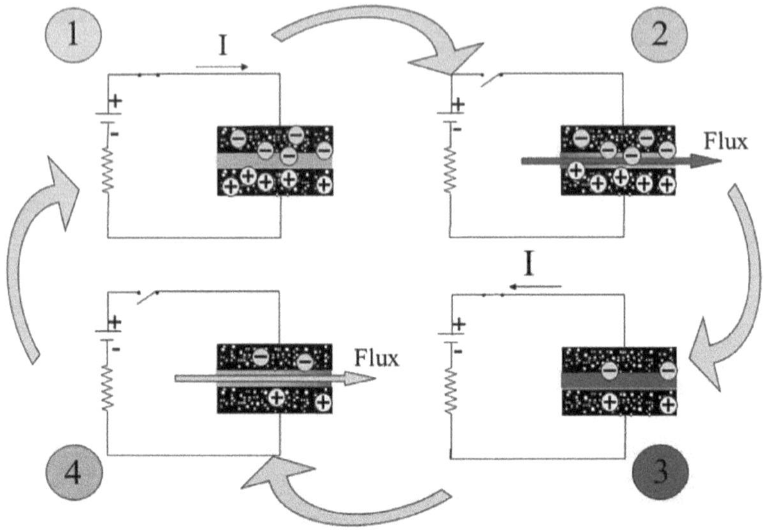

FIG. 6.6 Cell processes during every step of the DLPE cycle. 1: Cell charged in cold water; 2: hot water in; 3: electrode discharge; 4: cold water in. *(Reproduced with permission from Ahualli SA, Fernández MM, Iglesias GR, Delgado ÁV, Jimenez ML. Temperature effects on energy production by salinity exchange. Environ Sci Technol 2014;48:12378–85. Copyright 2014 American Chemical Society.)*

same potential difference will be established between both electrodes. Each of them will acquire a surface charge equal to σ_{min} in absolute value, and a potential difference $|\Psi_S| = |V|/2$ with respect to the solution in the space between them. Then, in open circuit, cold and hot waters are interchanged (step 2 in Fig. 6.6). As a consequence of the temperature increase, the permittivity decreases and so does the EDL capacitance. Since the circuit is open, this provokes an increase of the electric potential to $|\Psi_S| + |\Delta\Psi_S|$ at the electrode-solution interface. In order to use this increase in potential, we next discharge the electrodes (step 3) over the external source. Since the potential of the electrodes is larger, some charge will spontaneously migrate to the battery at a larger potential, resulting in a positive energy balance. A new equilibrium is attained, and the potential returns to the initial value but with a difference in surface charge σ_{max}. For closing the cycle, we exchange hot for cold water in open circuit (step 4). Finally, we again connect the battery to the cell filled with cold water (step 1) and return to the initial potential and charge density values.

The environmental conditions where this phenomenon can be used for energy harvesting include, for instance, the mixing of river and (colder) ocean waters (mostly if the river water has been used for refrigeration), where we could have the additional beneficial effect of salinity gradients on top of temperature ones.

Exploiting the temperature effect on energy production systems is an idea present in other techniques. For instance, Sales [22] has recently proposed to use the so-called *thermal membrane potential*: an electric potential is generated when hot and cold waters are contacted with anion and cation exchange membranes, respectively. That potential can be used in energy production by charging carbon electrodes in contact with the membranes, and discharging them through the external circuit. This is a modification of the capacitive mixing procedure known as CDP, as discussed in Chapter 7.

In this section, we show some experimental results concerning the implementation of CDLE with temperature differences between the two solutions used in the exchange process [8]. In Fig. 6.7A we show an example of the voltage between electrodes in a DLPE cycle, using 20 mM NaCl solutions. Note the steps of the capmix cycle when hot and cool waters are exchanged. It must be pointed out that, considering that the thermoelectric power of copper is 4 μV/K, the perturbation due to the Seebeck effect is negligible, and will not affect the readings shown.

Although feasible, the DLPE cycles produce a much lower energy than CDLE. In Fig. 6.7B successive cycles like that in Fig. 6.7A are shown in comparison with CDLE cycles. We can observe that the values of V_{cell} are lower in DLPE than in CDLE cycles, for otherwise identical charging conditions. Note that because the extracted work depends on both the charge exchanged with the external voltage source and the potential jump under solution exchange, and both are lower in DLPE, we can predict a reduced extracted work with the latter technique.

A way of improving the results was proposed in [8], and it regards the use of both at the same time. Data in Fig. 6.8 prove that properly combining

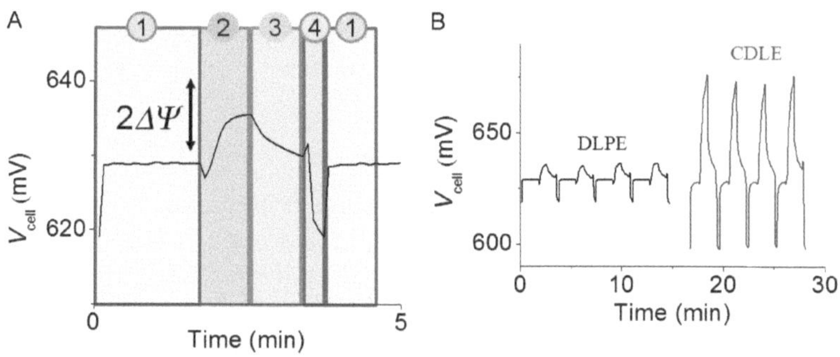

FIG. 6.7 Experimental results of the cell potential, V_{cell}, as a function of time. (A) A single DLPE cycle. $|\Delta\Psi|$ is the voltage jump in each electrode. (B) Successive cycles of DLPE (*left*) and CDLE (*right*). Charging voltage $V = 630$ mV; solution used in DLPE: 20 mM NaCl at 25 and 50°C; exchanged solutions in CDLE: 20 and 600 mM NaCl at 25°C. *(Reproduced with permission from Ahualli SA, Fernández MM, Iglesias GR, Delgado ÁV, Jimenez ML. Temperature effects on energy production by salinity exchange. Environ Sci Technol 2014;48:12378–85. Copyright 2014 American Chemical Society.)*

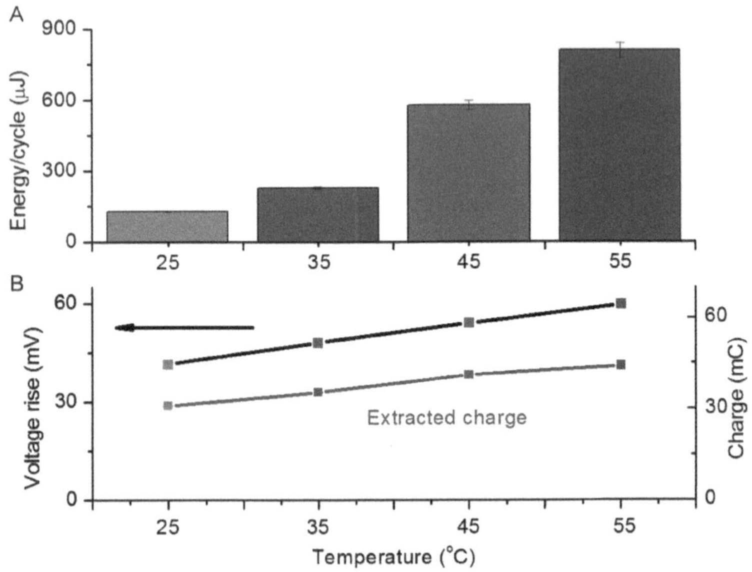

FIG. 6.8 Experimental results for CDLE + DLPE cycles. Charging solution: NaCl 500 mM at 25°C; discharging solutions: NaCl 20 mM at the indicated temperatures and 420 mV charging voltage. (A) Extracted energy values per cycle; (B) voltage rise and extracted charge for these same cases. Sample: SR51 film.

temperature and salinity differences makes it possible to maximize the voltage rise in the mixed CDLE + DLPE technique. As Fig. 6.8B indicates, if 500 mM is the concentration of the seawater at 25°C both the extracted charge of the cycle and the voltage rise increase with the temperature difference, and this brings about a significant rise in energy obtained (the increase can be fourfold for a 45°C temperature difference). This confirms the feasibility of the mixed technique in easily achieved conditions in practice.

6.6 Carbon Materials Requirements

In the capmix techniques, the need for exchange brings about the requirement that ions in the EDLs should be able to go in and out of them and they will most probably be hydrated, at least partially. The hydrophilicity of the carbon used for the electrode preparation is hence an important issue. In order to investigate this, we prepared two kinds of carbon particles, named SC-1 and SC-2 hereafter, with almost identical pore size distribution (Fig. 6.9A), but different wettabilities. In such samples any differences between CDLE performances must stem from surface properties.

As usual, the wettability was determined by measuring the contact angles of water on the carbon films. Both the advancing (θ_a) and receding (θ_r) angles were determined at 25°C. The difference between the two angles is called *contact angle hysteresis* [23] and an approach has been proposed relating both

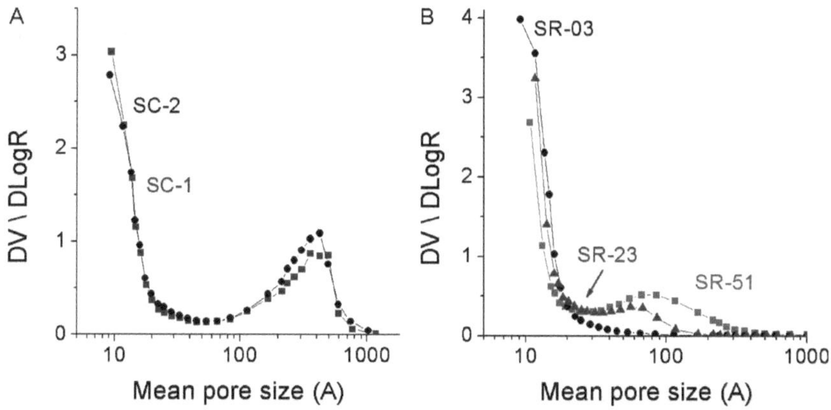

FIG. 6.9 Specific pore size distribution for samples SC-1 and SC-2 (A) and samples SR-03, SR-23, and SR-51 (B). *(Reproduced with permission from Iglesias GR, Fernández MM, Ahualli S, Jiménez ML, Kozynchenko OP, Delgado AV. Materials selection for optimum energy production by double layer expansion methods. J Power Sources 2014;261:371–7. Copyright 2014 Elsevier Ltd.)*

angles to the surface free energy of the solid, γ_s, which is the useful quantity for the study [23, 24]. This can be done as follows:

$$\gamma_s = \frac{\gamma_L (1 + \cos\theta_a)^2}{2 + \cos\theta_a + \cos\theta_r} \tag{6.2}$$

where γ_L is the liquid surface tension, water in our case (γ_L = 72.8 mN/m). The advancing contact angle was measured after depositing 6 ± 2μL droplets on the surface, and for the receding contact angle this volume was reduced to 4 ± 2μL. One example is shown in Fig. 6.10.

The results are presented in Table 6.3, which shows that sample SC-1 is very hydrophobic whereas carbon SC-2 is, on the contrary, characterized by hydrophilic behavior, and the SR samples are somewhere in between and rather similar to each other.

Fisrt we can explain the results obtained when SC-1 and SC-2 samples are used in CDLE cycles, as shown in Table 6.4. The voltage rise in SC-1 is almost half that is for SC-2 for otherwise identical conditions. In addition, the results in Table 6.4 show that extracting power is only possible for SC-2 sample demonstrating that hydrophobicity hinders the required ion exchange.

Let us now consider the role of pore size distribution, a property that can be expected to be essential: we need a certain fraction of macro- or mesopores as reservoirs and paths for ions, and a sufficient number of micropores, contributing to the high surface area required. In addition, the ratio between the ion and pore sizes determines the ease with which the ions can diffuse inside, and even if they are hydrated or (at least, partially) dehydrated [25]. Furthermore, the amount of charge transferred is limited by the possibility of EDL saturation because of the finite volume of counterions (Chapter 5). All these

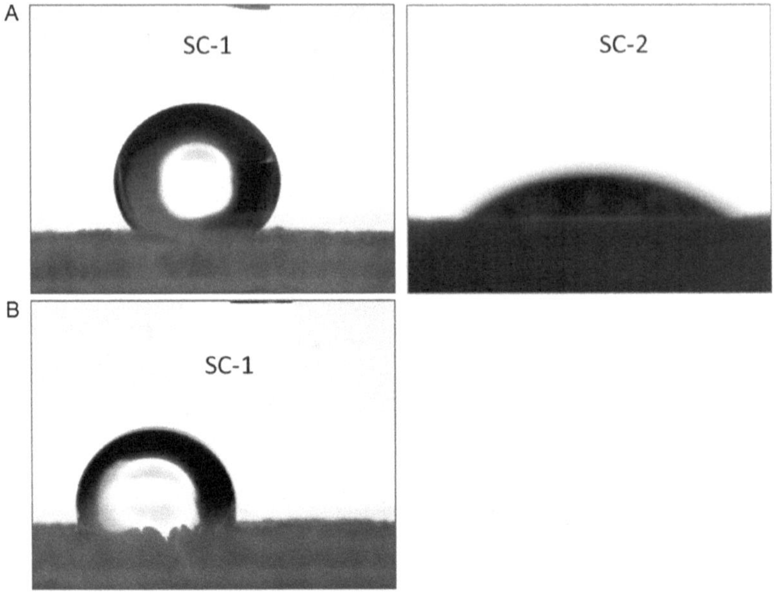

FIG. 6.10 Droplets formed for advancing (A) and receding (B) contact angle measurements for carbons SC-1 and SC-2. The receding contact angle could only be measured for the former, and it was taken as zero for SC-2.

TABLE 6.3 Advancing and Receding Contact Angles of Water on Samples SC-1, SC-2, SR03, SR-23, and SR-51

Sample	θ_a (degrees)	θ_r (degrees)	γ_s (mJ/m^2)
SC-1	125 ± 3	84 ± 7	8.5 ± 1.6
SC-2	29 ± 5	–	66.0 ± 2.1
SR-03	86 ± 5	49 ± 2	31 ± 4
SR-23	110 ± 2	74 ± 8	16 ± 1
SR-51	107 ± 3	61 ± 4	17 ± 2

Notes: The receding angle in SC-2 was not significantly distinct from zero. The surface free energy γ_s was calculated from Eq. (6.2).

TABLE 6.4 Voltage Rise and Extracted Power in Samples SC-1 and SC-2

Sample	Voltage Rise (mV)	Extracted Power (mW/m^2)
SC-1	24	–
SC-2	50	0.4

aspects may be determinant in the resulting kinetics (associated with the rate of ion adsorption-desorption in the EDL) and energy production (as the voltage rise and charge transferred will be reduced if ion exchange is incomplete). Even if exchange is not required, as in supercapacitor technology [25, 26], the pore size distribution is considered as the most important design parameter, taking into account the relationship between microstructure and ion accessibility to the pores.

The samples used for this study are described in Fig. 6.9B. Some results are displayed in Fig. 6.11. As theoretically expected [3], although the average pore diameter has negligible effect on the voltage rise (Fig. 6.11A), the larger quantity of pores in 1-nm size range produces a consistently higher delivered and extracted charge (22 mC in SR-03, to be compared to 9 mC in SR-51, for instance, in the extracted charge). Nevertheless, Fig. 6.11A demonstrates that the difference between delivered and extracted charges is also larger in case of predominantly small pore sizes. This can be observed in Fig. 6.11B where cell voltage is represented versus measured charge. The separations between successive cycles are clearly larger for samples SR-03 and SR-23. Interestingly, SR-51 contains a significant number of pores in the 5–10 nm range, so that charge leakage reduction can be explained by assuming that the presence of the larger pore fraction makes it possible that salty and fresh solutions are easily exchanged, leading to a favorable diffusion of counterions in and out of the micropores. Indeed, SR51 was the only sample where power extraction was possible, achieving 1 mW/m^2 [6].

6.7 Salinity Difference Effects

The research in CDLE so far tends to focus on the natural process of mixing river water with seawater in river mouths, as the main source of blue energy [4]. In recent years attention has also been paid to sources with other salinity differences, especially in dry areas. For instance, concentrated brine water

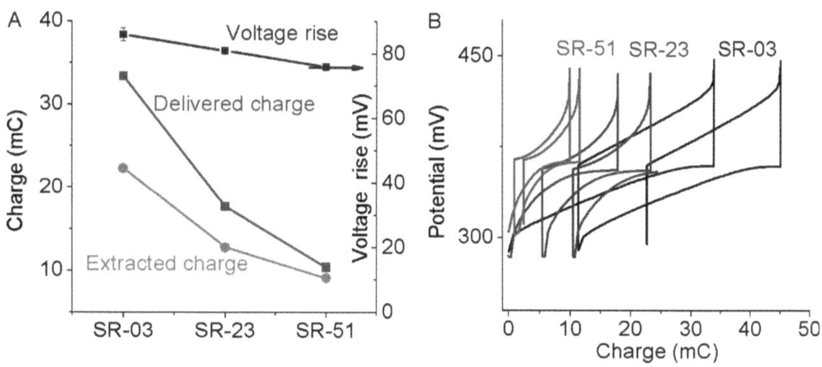

FIG. 6.11 (A) Voltage rise, extracted, and delivered charge for samples SR-03, SR-23, and SR-51. (B) Voltage versus charge CDLE cycles representation for the same samples.

from desalination plants can be paired with waste water from a treatment facility, reducing desalination energy cost [27] and additionally reducing its environmental impact. In this section, we propose the study of the effect of exchanging solutions on CDLE cycles with concentrations different from the commonly used river (20 mM) and sea (500 mM) waters.

Theoretical predictions [3] have shown the interest of using larger salinity differences. Fig. 6.12A represents surface charge density as a function of surface potential for four ionic concentrations: 1 M, 500 mM, 300 mM, and 20 mM of sodium chloride. Note that when the concentration is over 300 mM, the curves approach each other, as a consequence of the finite ion volume producing a less efficient screening of the pore wall surface charge, and hence show less steep potential decays and lower surface charge increase with potential. In summary, the potential rise and cycle charge do not increase linearly with the salinity gradient.

As is observed, when concentration increases over 300 mM, the curves seems to be closer. This is an important prediction of our model and it is due to the finite ion size. Although concentration increases, an unlimited counterion accumulation close to the surface is prevented. This leads to a less efficient screening of the surface charge and hence to flatter potential profiles and lower surface charge increase with surface potential. This produces the effect that voltage rise and cycle charge do not increase linearly with the salinity difference. Fig. 6.12B indicates that the largest cycle charge and voltage rise corresponds to the highest concentration difference (20 mV vs. 1 M).

Experimental results shown in Fig. 6.13 confirm the model predictions: Fig. 6.13A shows the same tendency of the voltage rise and the extracted charge as Fig. 6.12B. However, the experimental values are quite lower than the predictions, due to the already mentioned current leakage. Fig. 6.13B indicates furthermore that positive extracted power is only found for the three first cases (20 mV vs. 1 M, 500 mM, and 300 mM) reaching its maximum at 1.4 mW/m^2 for the largest salinity difference [6].

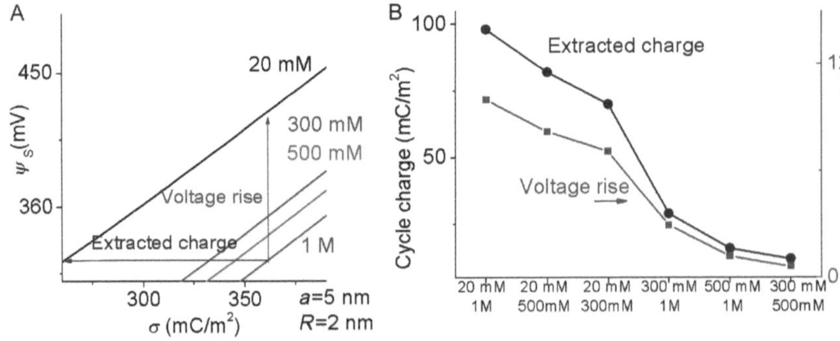

FIG. 6.12 (A) Surface charge density versus surface potential for four ionic concentrations: 1 M, 500 mM, 300 mM, and 20 mM. (B) Voltage rise and cycle charge for six different pairs of exchanged solutions. Sample: SR51 film.

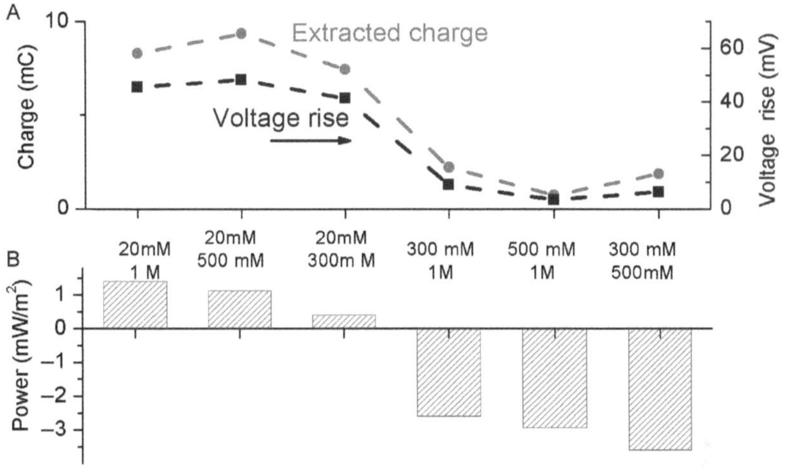

FIG. 6.13 (A) Voltage rise and extracted charge for six different pairs of exchanged solutions. (B) Extracted power for these same cases. Sample: SR51 film.

These results examine the promising implementation of CDLE for different salinity sources. CDLE performance can be maximized by enlarging the salinity difference among the solutions exchanged. This fact gives the opportunity to other water solutions, in particular, highly contaminating brine solutions, which might be a low-cost feed.

6.8 Stackings of Capmix Cells

In order to implement capmix techniques in an electronic circuit, higher voltages than those reported previously are required, and this can only be achieved by associating individual cells, as already proposed for other capmix techniques [28–30] and we now consider some possible ways to do this [10], as, for example, in Fig. 6.2A. The comparison between voltage and current oscillations along the cycles, obtained with one, two, and three cells, is presented in Fig. 6.14. Note that the voltage needed for charging is roughly three times larger when we have three cells in series as compared to just one; however, the currents measured during charge and discharge are very similar in all the cases, indicating that the internal resistance also increases at the same rate. Regarding the energy per cycle, the calculated values are 330, 620, and 400 mJ/m^2, certainly lower than expected.

A different stack can be designed in order to ascertain the reasons for these apparently anomalous results of the cell association, denominated here stack 2 (Fig. 6.2B), whereby each cell can be independently connected to the external circuit and measurements can be taken separately. The results are of the kind plotted in Fig. 6.15, containing the voltage of each electrode pair (numbered as in Fig. 6.2B) and the total current during successive cycles, after charging with 1.48 V between the outermost electrodes.

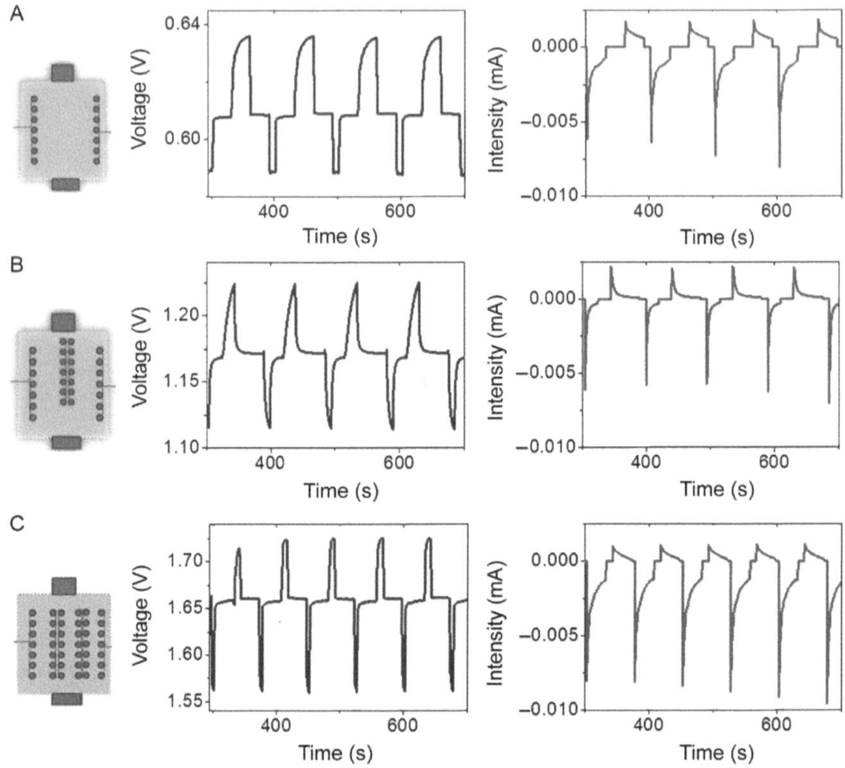

FIG. 6.14 Voltage and current cycles in the CDLE experiments with one (A), two (B), and three (C) cells connected in series. The charging voltages were 0.605 V (A), 1.17 V (B), and 1.66 V (C). *(Reproduced with permission from Iglesias GR, Ahualli S, Fernández MM, Jiménez ML, Delgado ÁV. Stacking of capacitive cells for electrical energy production by salinity exchange. J Power Sources 2016;318:283–90. Copyright 2016 Elsevier Ltd.)*

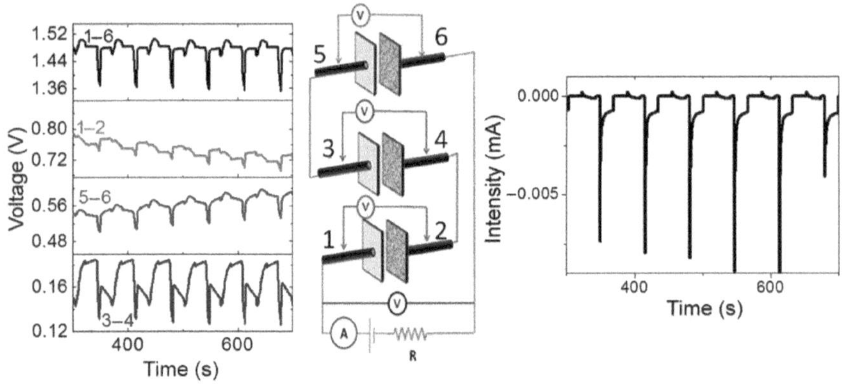

FIG. 6.15 Voltage (*left*) and current (*right*) oscillations during the capmix cycles. The *left panel* shows the oscillations of the voltage difference between the indicated electrode pairs. *(Reproduced with permission from Iglesias GR, Ahualli S, Fernández MM, Jiménez ML, Delgado ÁV. Stacking of capacitive cells for electrical energy production by salinity exchange. J Power Sources 2016;318:283–90. Copyright 2016 Elsevier Ltd.)*

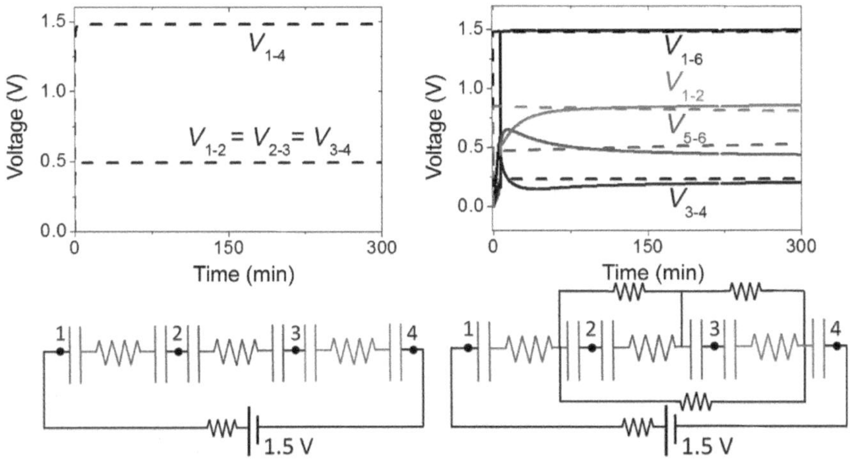

FIG. 6.16 Time evolution of the voltage of individual capacitors in a series association. *Left*: Conductive connections exist only between neighbor plates. *Right*: Experimental charging data in stack 2 (*solid lines*), and simulations (*dotted lines*) when additional connections are established through external resistors (*bottom*). The parameters used for the simulation are: C = 1 F; R_L = 5 Ω; R = 100 Ω; R_1 = 50 Ω; R_2 = 200 Ω; R_3 = 500 Ω. *(Reproduced with permission from Iglesias GR, Ahualli S, Fernández MM, Jiménez ML, Delgado ÁV. Stacking of capacitive cells for electrical energy production by salinity exchange. J Power Sources 2016;318:283–90. Copyright 2016 Elsevier Ltd.)*

As observed, the initial charging voltages are different for each cell of the association (Fig. 6.15, left), a fact that can be explained considering the circuit simulation of the charging of a series of capacitors (using LTspice IV, Linear Technology, Norwood, MA). Two situations are considered, namely, an ideal one in which only connections with adjacent cells are assumed (Fig. 6.16, top), and a more realistic one in which additional connections are established, simulating those in the actual cell stack through the electrolyte solution (Fig. 6.16, bottom). In this situation, the measured voltage of every cell is in good agreement with the simulation results. This simulation demonstrates that the additional conduction between plates of different cells leaves the innermost one (not directly connected to the external circuit) inoperative. It can be concluded that it is just the capacitors in direct contact with the external source that can contribute to the overall voltage rise, contrary to the negligible role of the internal cell. As illustrated in Fig. 6.15, the voltage rises (Table 6.5) decrease when the effective charging voltage is above 300 mV, roughly, a confirmation of the previously mentioned existence of an optimum charging voltage. Any cell not reaching this voltage in the stack will be essentially inoperative for energy production. This is proposed to be a limitation of any series connection of capmix CDLE cells.

These results can be considered of great significance for the implementation of CDLE in situations closer to practical: the fact that the inner cell is not

TABLE 6.5 Charging and Rise Voltages for the Different Electrode Pairs Connected in Series as in Fig. 6.2B

Pair	Charging Voltage (mV)	Voltage Rise (mV)
1–2	0.761	−5
3–4	0.145	+30
5–6	0.558	+15
1–6	1.460	+40

contributing and that we cannot guarantee that all the cells are working in their optimum charging voltage limits the validity of the intuitively simple association proposed at the beginning of this section.

6.9 Conclusions

In this chapter, we have described the different issues associated with appropriate CDLE implementation in order to maximize the power and energy extracted. We have demonstrated that improvements are expected to come from: minimization of the charge leakage, optimum selection of the working potential, and adequate charging process to maintain energy production cycles stable and reproducible.

Extracted energy from use of the CDLE technique can also be improved if the temperature of the solutions is suitably chosen. The results discussed show that the technique based on the dependence of the electric double layer capacitance on temperature can be used as a new approach to the extraction of electrical energy from thermal gradients without the need for electromechanical converters. However, considering that the CDLE + DLPE technique is just starting to be implemented, it is reasonable to expect significant improvements in efficiency and power density when the technology is further developed. In addition, it can be used in combination with other well-known sources of renewable energy, such as solar thermal plants, geothermal extraction, or even home solar collectors as a source of heat during the long periods of inactivity.

Subsequently, we have shown that the average pore size of activated carbon particles, in relation to the EDL thickness and the ion diameter is, as expected, a determinant quantity. Optimum CDLE results can be obtained if the carbon used in the electrodes has a significant fraction of mesopore population, which behaves as a sort of solution reservoir allowing easy water exchange. Additionally, results would be improved if hydrophilized activated carbon materials are employed: this might look counterintuitive, since carbon oxidation can be favored in such a case. However, the fact that hydrated ions must be cyclically exchanged with the contacting solutions compensates for this possible drawback. Moreover, this chapter would not be complete without the inclusion in this research of other less common salinity sources that could improve extracted power as well as its positive environmental impact, as is the case for saturated water solutions, which mean an important potential risk to aquatic resources.

Finally, searching for increased energy and power production in CDLE techniques, we investigate the possibility of associating individual cells (pairs of porous electrodes) in series. It is found that the voltage rise when salt water is substituted by fresh water is roughly a multiple of the rises of individual cells, while the current is the same in all cases. An analysis performed on individual cells demonstrates that the conductive connection between nonadjacent units due to the permeating electrolyte solution leaves only the outermost cells as actually working ones.

References

[1] Delgado A. Interfacial electrokinetics and electrophoresis. Surfactant science series, vol. 106. New York: CRC Press; 2002.

[2] Wall S. The history of electrokinetic phenomena. Curr Opin Colloid Interface Sci 2010;15(3):119–24.

[3] Jiménez ML, Fernández MM, Ahualli S, Iglesias G, Delgado AV. Predictions of the maximum energy extracted from salinity exchange inside porous electrodes. J Colloid Interface Sci 2013;402:340–9.

[4] Ramon GZ, Feinberg BJ, Hoek EMV. Membrane-based production of salinity-gradient power. Energy Environ Sci 2011;4:4423–34.

[5] Yip NY, Elimelech M. Comparison of energy efficiency and power density in pressure retarded osmosis and reverse electrodialysis. Environ Sci Technol 2014;48(18):11002–12.

[6] Fernández MM. Analysis of energy production by salinity exchange in porous electrodes [Ph.D. thesis]. Spain: Univ. Granada; 2017.

[7] Iglesias GR, Fernández MM, Ahualli S, Jiménez ML, Kozynchenko OP, Delgado AV. Materials selection for optimum energy production by double layer expansion methods. J Power Sources 2014;261:371–7.

[8] Ahualli SA, Fernández MM, Iglesias GR, Delgado ÁV, Jimenez ML. Temperature effects on energy production by salinity exchange. Environ Sci Technol 2014;48:12378–85.

[9] Delgado AV, Ahualli S, Fernández MM, González MA, Iglesias GR, Vivo-Vilches JF, et al. Geometrical properties of materials for energy production by salinity exchange. Environ Chem 2017;14(5):279–87.

[10] Iglesias GR, Ahualli S, Fernández MM, Jiménez ML, Delgado ÁV. Stacking of capacitive cells for electrical energy production by salinity exchange. J Power Sources 2016;318: 283–90.

[11] Iglesias GR, Ahualli S, Delgado AV, Jiménez ML, González-Caballero F. A device for the measurement of energy by ionic exchange. EU Patent PCT/ES2014/070722. University of Granada; 2014.

[12] Conway BE, Pell WG, Liu TC. Diagnostic analyses for mechanisms of self-discharge of electrochemical capacitors and batteries. J Power Sources 1997;65(1):53–9.

[13] Conway BE. Electrochemical supercapacitors: scientific fundamentals and technological applications. New York: Kluwer Academic; 1999.

[14] Biesheuvel PM, Fu YQ, Bazant MZ. Diffuse charge and faradaic reactions in porous electrodes. Phys Rev E 2011;83(6):061507.

[15] Liu F, Schaetzle O, Sales BB, Saakes M, Buisman CJN, Hamelers HVM. Effect of additional charging and current density on the performance of capacitive energy extraction based on Donnan potential. Energy Environ Sci 2012;5:8642–50.

[16] Tennison S, Kozynchenko O, Strelko V, Blackburn A. A US Pat US20040024074 A1; 2004.

[17] Masliyah JH, Bhattacharjee S. Electrokinetic and colloid transport phenomena. Hoboken (NJ): Wiley-Interscience; 2006.

[18] Brogioli D, Zhao R, Biesheuvel PM. A prototype cell for extracting energy from a water salinity difference by means of double layer expansion in nanoporous carbon electrodes. Energy Environ Sci 2011;4(3):772–7.

[19] Biesheuvel PM, Bazant MZ. Nonlinear dynamics of capacitive charging and desalination by porous electrodes. Phys Rev E 2010;81:031502.

[20] Kaus M, Kowal J, Sauer DU. Modelling the effects of charge redistribution during self-discharge of supercapacitors. Electrochim Acta 2010;55(25):7516–23.

[21] Kowal J, Avaroglu E, Chamekh F, S'enfelds A, Thien T, Wijaya D, et al. Detailed analysis of the self-discharge of supercapacitors. J Power Sources 2011;196(1):573–9.

[22] Sales BB. Capacitive technology for energy extraction from chemical potential differences [Ph.D. thesis]. The Netherlands: Univ. Wageningen; 2013.

[23] Chibowski E. Wettability and adhesion. In: 2nd international symposium on contact angle, Newark, NJ, Jun 21–23; 2000.

[24] Chibowski E. On some relations between advancing, receding and young's contact angles. Adv Colloid Interface Sci 2007;133(1):51–9.

[25] Qu QT, Wang B, Yang LC, Shi Y, Tian S, Wu YP. Study on electrochemical performance of activated carbon in aqueous Li_2SO_4, Na_2SO_4 and K_2SO_4 electrolytes. Electrochem Commun 2008;10(10):1652–5.

[26] Endo M, Takeda T, Kim YJ, Koshiba K, Ishii K. High power electric double layer capacitor (EDLC's); from operating principle to pore size control in advanced activated carbons. Carbon Sci 2003;1(3–4):117–28.

[27] Achilli A, Prante JL, Hancock NT, Maxwell EB, Childress AE. Experimental results from ro-pro: a next generation system for low-energy desalination. Environ Sci Technol 2014;48:6437–43.

[28] Burheim OS, Liu F, Sales BB, Schaetzle O, Buisman CJN, Hamelers HVM. Faster time response by the use of wire electrodes in capacitive salinity gradient energy systems. J Phys Chem C 2012;116(36):19203–10.

[29] Sales BB, Burheim OS, Liu F, Schaetzle O, Buisman CJN, Hamelers HVM. Impact of wire geometry in energy extraction from salinity differences using capacitive technology. Environ Sci Technol 2012;46(21):12203–8.

[30] Liu F, Donkers TFW, Wagterveld RM, Schaetzle O, Saakes M, Buisman CJN, et al. Parallel up-scaling of capacitive mixing (capmix) system enhances the specific performance. Electrochim Acta 2016;187:104–12.

Section D

Capacitive Energy Extraction by Donnan Potential (CDP)

Chapter 7

CDP: Fundamentals of the Method

Silvia Ahualli, María M. Fernández, Ángel V. Delgado and Guillermo R. Iglesias
Department of Applied Physics, University of Granada, Granada, Spain

7.1 Introduction

In the capacitive double layer expansion (CDLE) capacitive mixing (CapMix) technology dealt with in previous chapters, the accumulation of counterions on the electrodes is achieved by application of a potential difference between them by connection to a constant-voltage or constant-current power source. In the event of calculating costs or performances, the electrical work performed by the power source upon charging must be subtracted from the energy harvested during desorption. In addition, the self-discharge of the cell is a persistent drawback that has been faced also by the supercapacitor technologies. Electrical double layers (EDLs) exhibit significant rates of self-discharge mainly as a consequence of the ease with which the charge on the electrode can be affected by impurity reactions. Although all details are not known, it is admitted that an ideal, nonleaky capacitor (ideally polarizable electrodes) will show no discharge phenomena. More specifically, possible mechanisms have been offered by Conway [1, 2], and these include, in addition to the faradaic processes previously mentioned, overcharging beyond the limit of the potential decomposition of the electrolyte or existence of ohmic paths due to imperfect sealing or undesired contacts between opposing electrodes.

In 2010, Sales et al. [3] proposed an entirely different approach. Their main innovation consisted of substituting ion exchange membranes (cationic and anionic, respectively, on each electrode) sandwiched between the porous carbon electrodes (Fig. 7.1) for the external power source. The concentration gradient between the carbon wall and the solution in the midchannel produces a membrane potential in the presence of salty water in the channel. The membrane allows the passage of only one type of ions (assuming for the moment 100% selectivity) towards its respective counterelectrode, as produced by the power source in the CDLE cell. If a connection is established between them

Interface Science and Technology, Vol. 24. https://doi.org/10.1016/B978-0-12-811370-7.00007-3

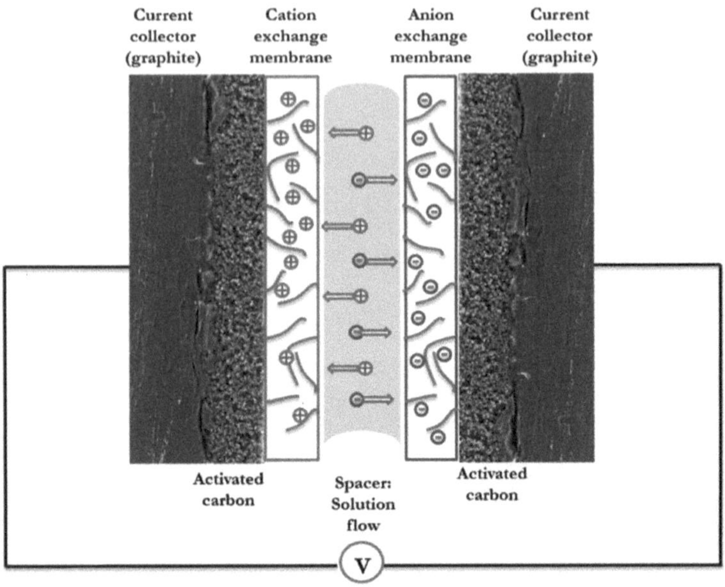

FIG. 7.1 Schematics of the CDP CapMix cell.

through an external circuit, a current will flow, and a subsequent accumulation of electrons will take place close to the adsorbed cations on one electrode, and a depletion close to the adsorbed anions on the other. Conversely, if a low-salt solution (*river* water) is pumped between or through the electrodes, the corresponding ions migrate out of the latter and into the channel. The current will eventually flow in the opposite direction, but it is again usable for power generation.

In this chapter we will first consider the fundamentals of the method. Theoretical explanations of the processes will be compared to experimental data, with attention to both the steady state and the kinetics of the capacitive energy extraction based on the Donnan potential (CDP) approach. Consideration will also be given to proposed improvements, the use of wire-shaped electrodes in particular, and to the performance in mixed solutions. Finally, some estimations of efficiency compared to that achieved in other approaches will also be offered.

7.2 Details of the CDP Method

Let us first consider the reason for the denomination given to this technique. Imagine the anion exchange membrane, built of fixed cationic groups, in contact with the sea water. Because this membrane is formed by fixed ionizable groups, the transport of counterions is favored beyond diffusion until the electric field created by their excess prevents further transport. The potential difference between the interior of the membrane and the external solution, associated with such field, is called the Donnan potential Ψ_{Don}. If the membrane is perfectly selective, and it is surrounded by different solutions on both sides, the membrane potential will be the difference between the Donnan potentials on each side, in our case the bulk solution and the electrode (monovalent electrolyte) [4]:

$$\Psi_{Don} = \frac{k_B T}{e} \ln \frac{c_{bulk}}{c_{elec}} \tag{7.1}$$

If we assume that the initial state of the cell is fixed at zero potential difference and bathed by the salt solution (500 mM NaCl, say), then

$$\begin{aligned} c_{bulk} &= c_{salt} \\ c_{elec} &\simeq c_{salt} \end{aligned} \tag{7.2}$$

and the Donnan potential will be close to zero on both electrodes. When a fresh water (20 mM) solution is pumped in, the condition (7.2) now changes to

$$\begin{aligned} c_{bulk} &= c_{fresh} \\ c_{elec} &\simeq c_{salt} \end{aligned} \tag{7.3}$$

and a potential difference of the order of $2k_BT/e \times \ln(500/20) \cong 145$ mV will be generated between the electrodes. The cycle is typically implemented in four steps, much like the CDLE one (Fig. 7.2):

(i) Electrodes short-circuited immersed in salt water. Zero potential difference between the electrodes. No current.
(ii) Open circuit, fresh water pumped in. Membrane potential generated. No current.

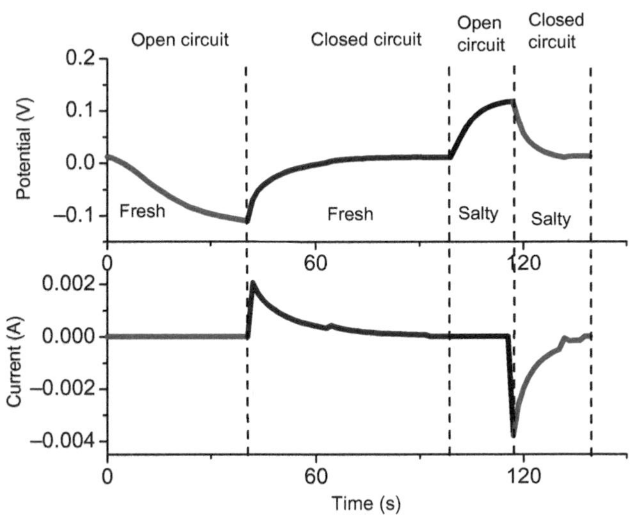

FIG. 7.2 Typical steps in a CDP cycle.

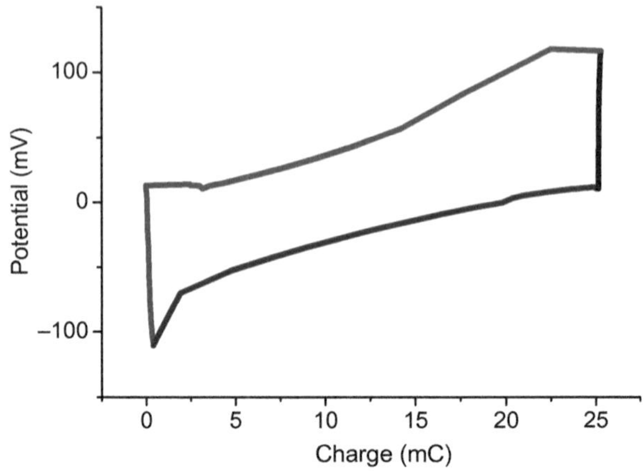

FIG. 7.3 *V-Q* cycle: relation between potential and charge in the CDP cycle shown in Fig. 7.2.

(iii) Closed circuit through the load resistor. The potential difference evolves towards zero while current is passing through the circuit. At the end, the Donnan potential difference is compensated for by the potential drop in the EDLs of the micropores. The potential difference between the electrodes drops to zero, but an amount of charge has been transferred from one to the other through the external circuit.

(iv) Open circuit, salt water in. The Donnan potential difference is reduced.

(v) Closed circuit. The accumulated charge is transferred back.

The energy that can be harvested per cycle can be evaluated by calculating, by integration of the current, the charge transferred in the closed-circuit steps and plotting the cell potential vs. the charge (Fig. 7.3). If no leakage is present, the area of the cycle will represent the energy extracted. Nevertheless, leakage cannot be ruled out [3]: this is the flow of solutions through the membranes into the electrodes, whereby the concentration difference between the spacer and the latter is reduced, leading to lower potential differences.

7.3 Theory: Steady State

A complete account of the time evolution of the potential and charge transferred in a CDP experiment was reported by Sales et al. [3]. This and other approaches searching for a description of the kinetics of the phenomenon will be dealt with later in this chapter. Here we will limit ourselves to finding the steady-state situation regarding the cell potentials in the different stages of the process [5]. Our starting point is the simple Ohshima et al.'s model of an ion exchange membrane [6, 7]. The membrane is planar (x will be the direction perpendicular to it, and the only dimension of interest), and characterized by its concentration of fixed ionizable groups (positive/negative for an anion/cation exchange membrane), ρ_{mem}, its thickness, d_c, and its permselectivity, which will be considered ideal. Our target is the determination of the membrane

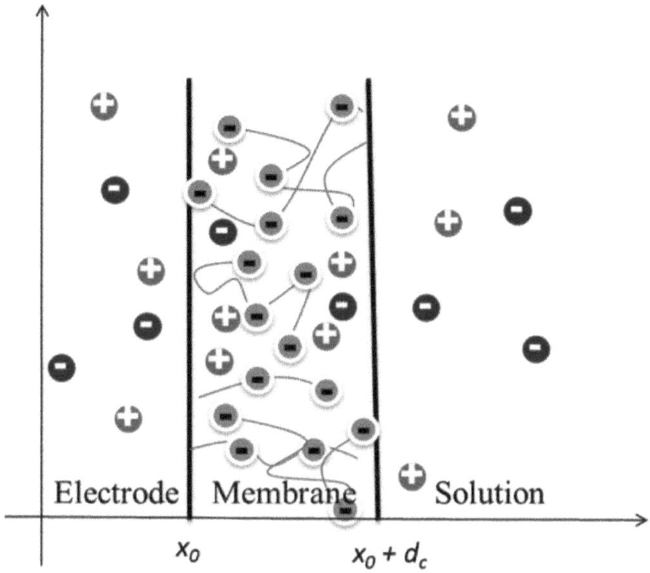

FIG. 7.4 Scheme of an ion (cation, in this case) exchange membrane. x_0 is the position of the left surface of the membrane, and x_0+d_c is that of the right wall. *Reproduced with permission from Ahualli S, et al. Effect of solution composition on the energy production by capacitive mixing in membrane-electrode assembly, J Phys Chem C 2014;118:15590. Copyright 2014 American Chemical Society.*

potential, that is, the potential difference between both faces of the membrane, for known ionic concentrations on the left and right baths, n_i^L, n_i^R $i=1, 2$, respectively. For simplicity, we will consider only two types of monovalent ions ($z_1=-z_2=1$; $n_1^{L,\ R}=n_2^{L,\ R}=n^{L,\ R}$), with equal diffusion coefficients, $D_1=D_2=D$, although more general cases are considered in [5]. Fig. 7.4 is a schematic representation.

We first consider the Poisson equation outside and inside the membrane. The electrical permittivity is $\varepsilon\varepsilon_0$ outside and $\varepsilon_m\varepsilon_0$ in the membrane:

$$\begin{aligned}\frac{d^2\Psi}{dx^2}&=-\frac{e(n_1-n_2)}{\varepsilon\varepsilon_0},\ \text{outside the membrane}\\ \frac{d^2\Psi}{dx^2}&=-\frac{e(n_1-n_2)}{\varepsilon_m\varepsilon_0}-\frac{\rho_{mem}}{\varepsilon_m\varepsilon_0},\ \text{inside the membrane}\end{aligned} \tag{7.4}$$

Following Ohshima, it will be assumed that the flux of both types of ions is constant inside the membrane:

$$\begin{aligned}J_i&=-D\frac{dn_i}{dx}-\frac{ez_in_i}{k_BT}\frac{d\Psi}{dx}=C_i,\quad i=1,2,\ \text{inside}\\ J_i&=0,\quad i=1,2,\ \text{outside}\end{aligned} \tag{7.5}$$

where e is the elementary charge and k_BT is the thermal energy.

We note that this system of equations has 15 unknowns (namely: $\Psi(x)$, $d\Psi/dx$, $n_1(x)$, $n_2(x)$ in each of the three regions (L, mem, R); the membrane potential $\Delta\Psi_m$; the constant fluxes C_1, C_2. The necessary boundary conditions come, first of all, from the continuity of the potential and the normal component of the displacement on the membrane surfaces:

$$\begin{aligned} \Psi\left(x_0^-\right) &= \Psi\left(x_0^+\right) \\ \Psi\left[(x_0+d_c)^-\right] &= \Psi\left[(x_0+d_c)^+\right] \\ \varepsilon \left.\frac{d\Psi}{dx}\right|_{x_0^-} &= \varepsilon_m \left.\frac{d\Psi}{dx}\right|_{x_0^+} \\ \varepsilon_m \left.\frac{d\Psi}{dx}\right|_{(x_0+d_c)^-} &= \varepsilon \left.\frac{d\Psi}{dx}\right|_{(x_0+d_c)^+} \end{aligned} \tag{7.6}$$

Also, the concentrations are assumed to be continuous at the electrode-membrane (x_0) and membrane-spacer channel ($x_0 + d_c$) interfaces:

$$\begin{aligned} n_i\left(x_0^-\right) &= n_i\left(x_0^+\right), \quad i=1,2 \\ n_i\left[(x_0+d_c)^-\right] &= n_i\left[(x_0+d_c)^+\right], \quad i=1,2 \end{aligned} \tag{7.7}$$

Finally, at large distance from the membrane interfaces, on both sides, the values of the potential and concentrations are specified:

$$\begin{aligned} &\Psi(x\to-\infty)=\Delta\Psi_m \\ &\Psi(x\to+\infty)=0 \\ &n_i(x\to-\infty)=n^L, \quad i=1,2 \\ &n_i(x\to+\infty)=n^R, \quad i=1,2 \end{aligned} \tag{7.8}$$

thus summing up to 14 conditions. The last one that is needed comes from the requirement that in steady state there is no electric current through the system:

$$e(J_1-J_2)=0 \tag{7.9}$$

The problem is solved using the routine BVP4C in Matlab. Fig. 7.5 shows the results for two values of ρ_{mem} (0.5 M and 1 M, or 48 and 96 MC/m^3, respectively), and two NaCl concentrations in the channel (0.5 and 0.02 M). Note that, when the concentrations on both sides of the membrane are different, a potential difference (the difference between the corresponding Donnan potentials) is established, which changes sign when the external solution changes from above to below that in the electrode. It is this potential that leads to the electronic current in the external circuit when the electrodes are connected through the external resistor.

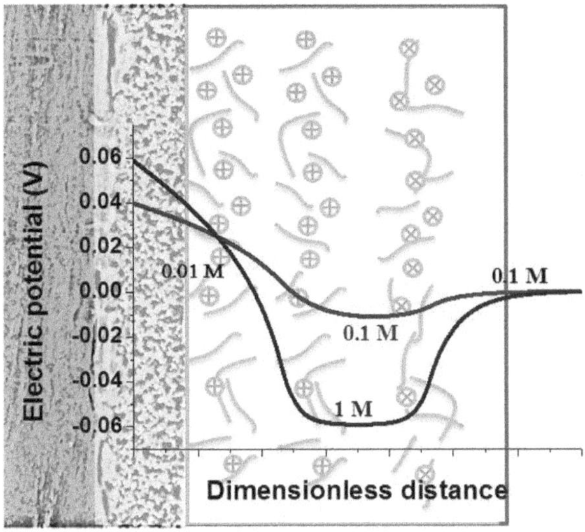

FIG. 7.5 Electrical potential profiles in the ion-exchange membrane and in the solution and electrode in contact with it. It is assumed that the membrane has charges of 1 M or 0.1 M, and that the concentration of the solution in the spacer channel is 0.1 M. The concentration inside the electrode is set at 0.01 M. The membrane thickness is 35 μm.

7.4 "Soft" Electrodes

A related approach to salinity gradient energy was proposed in [8, 9]. Fig. 7.6 illustrates the basics of this method. The carbon electrodes are coated with a layer of polyelectrolyte (a soft, solvent-permeable, highly charged layer) thick enough for a Donnan potential to be generated, similarly to the described behavior for membranes, but without the need for them [10–16]. The cationic and anionic polyelectrolyte coatings are assumed to be symmetrical, that is, they only differ in the sign of the fixed charges. At step 0 of Fig. 7.6A the electrodes are in open circuit, wetted by salt water, so that the Donnan potentials generated are Ψ_{Don}(salt) and $-\Psi_{Don}$(salt) (respectively, $\Delta\Psi_D^+$, $\Delta\Psi_D^-$ in Fig. 7.6B). The potential difference in the cell will be $\Delta V = 2\Psi_{Don}$(salt). Upon external connection (stage 1), a charge transfer will take place and EDLs are built on each electrode and a potential drop $\Delta\Psi_{EDL} = -\Psi_{Don}$(salt) will occur at the electrode/layer interface. If the circuit is open and fresh water is pumped in the channel (step 2), the potential will rise as the Donnan potentials of the subcells do, and $\Delta V = 2\Psi_{Don}(\text{fresh}) + 2\Delta\Psi_{EDL}(\text{fresh})$. If the electrodes are short-circuited again (stage 3), current will be transported and the potential difference will drop to zero, while the electrodes remain charged (notice the slope of the potential at the interfaces). The system is reset by allowing sea water in open circuit (step 4). The potential profile is shifted according to the Donnan potential changes, and finally the external connection brings the system to the starting point.

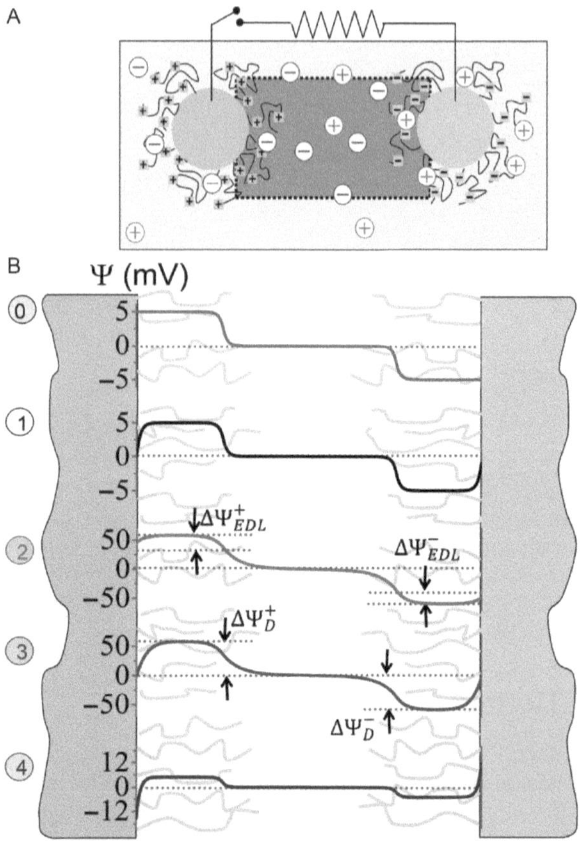

FIG. 7.6 (A) Two carbon particles are assumed to be coated by polyelectrolyte layers, thus becoming *soft* particles. (B) Two electrodes are built with these particles, coated respectively by anionic and cationic polyelectrolytes. The steady-state potential profiles shown correspond to: 0: open circuit, electrodes immersed in salt solution; 1: electrodes connected externally; 2: fresh water pumped in open circuit; 3: external connection; 4: salt water in, open circuit. The scales of the potential axes are different in the presented steps. *Reproduced from Ahualli S, et al. Polyelectrolyte-coated carbons used in the generation of blue energy from salinity differences, Phys Chem Chem Phys 2014;26:25241 with permission from the PCCP Owner Societies.*

This procedure resembles the CDP one, except for the different origin of the electrode potential: in CDP the membranes are in contact with solutions of different salinities on both sides (Fig. 7.4), that is, different Donnan potentials on each side, and a membrane potential difference between them. In the case of soft electrodes (SEs), the polyelectrolyte resides almost directly on the electrode particles, and the surface potentials concerned are just Donnan potentials.

The equation system governing the potential and ion concentration profiles is thus very similar to that described previously for CDP. Differences are to be noted, however. The problem is now solved for a single particle (which in fact mimics the entrance to micropores) from the ion reservoirs, that is, the macropores. Eq. (7.4) is solved assuming a Boltzmann distribution of ions:

$$n_i = n_i^{bulk} \exp\left(-z_i e \Psi / k_B T\right) \tag{7.10}$$

It is also assumed that the symmetry is spherical and the following boundary conditions apply (σ is the surface charge density):

$$\begin{aligned} \left.\frac{d\Psi}{dr}\right|_{core} &= -\frac{\sigma}{\varepsilon\varepsilon_0} \\ \left.\frac{d\Psi}{dr}\right|_{core} &= 0 \end{aligned} \tag{7.11}$$

In the boundary between the coating and the solution ($r = r_C$) there is continuity of both the potential and the normal displacement:

$$\begin{aligned} \Psi|_{r_C^-} &= \Psi_{r_C^+} \\ \left.\frac{d\Psi}{dr}\right|_{r_C^-} &= \left.\frac{d\Psi}{dr}\right|_{r_C^+} \end{aligned} \tag{7.12}$$

Solving the equations as before, it is possible to obtain a theoretical cycle, for selected values of the parameters of interest, as illustrated in Fig. 7.7. The area of the cycle would correspond to the energy available in the conditions shown, namely 3.1 J/m^2; this is a really large value, and later on in the chapter we will consider how much of it is feasible.

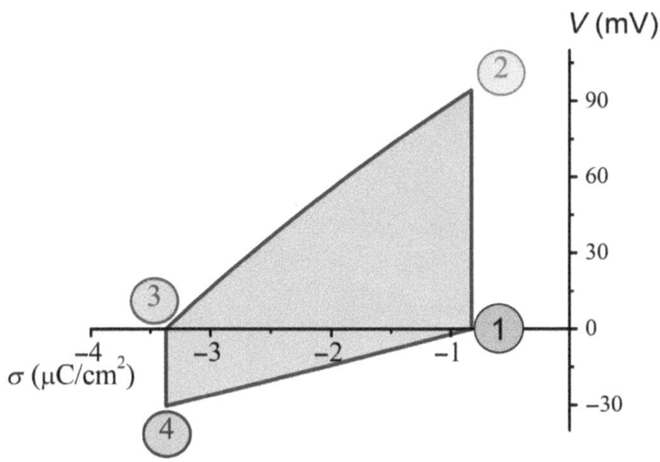

FIG. 7.7 Simulated SE cycle in the following conditions: salt solution, 500 mM NaCl, fresh solution 20 mM NaCl; polyelectrolyte fixed charges +6 and −3.8 MC/m^3. *Reproduced from Ahualli S, et al. Polyelectrolyte-coated carbons used in the generation of blue energy from salinity differences, Phys Chem Chem Phys 2014;26:25241 with permission from the PCCP Owner Societies.*

7.5 Experimental Implementation

The basics of the methods are relatively easy, and the implementation follows almost exactly the described schematics of the CDP and SE. One needs to deposit first of all the high-capacitance (activated) carbon electrodes on a substrate (platinum, titanium, very often graphite) acting as current collector and electrically contacted with the external circuit. The solution in the spacer channel and the active electrodes are separated by ion exchange membranes of the adequate polarity (see details in [3, 17]).

As an example, in Fig. 7.8 we depict the cell used in our experiments, consisting of two opposing circular electrodes 1 cm in radius, made out of a porous carbon with pore distribution as indicated in the figure. The membranes used were manufactured by Astom, Japan, under the trade name of Neosepta. As observed in Figs. 7.2 and 7.3, the cell potential changed by about 120 mV, when salt and fresh water solutions were pumped through the cell. The current in the external circuit, in turn, can reach some 3.5 mA, corresponding to an average charge of 25 mC transferred back and forth in each cycle, yielding a total of 0.014 mW per cycle or, as usually expressed, 23 mW/m^2, using the projected (external) area of the two electrodes (6.28×10^{-4} m^2).

The values of energy per cycle are certainly limited if we are restricted to working with just 100 cm^2 of electrode area. In order to increase the current available, and thus the energy per cycle and the overall power, it is required to multiply the number of active electrode/membrane couples, as illustrated in Fig. 7.9 [3].

A somewhat related methodology, also requiring the use of membranes, has recently been proposed by Zhan et al. [18], named *concentration capacitor* by the authors, and depicted in Fig. 7.10. Several innovations can be cited in relation with this design. First of all, the electrodes are built with graphene hydrogels (GH) [19, 20], produced in the form of self-maintained blocks, not needing adhesive, hence prone to be sliced and deposited directly

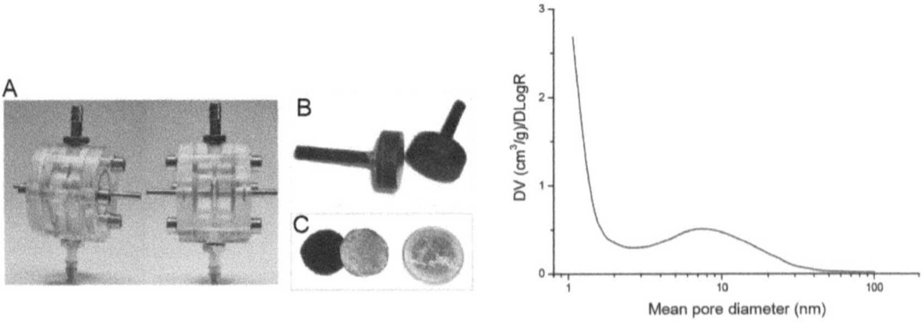

FIG. 7.8 Left: the cell used in our experiments with CDP; (A) main plastic body; (B) graphite current collectors; (C) graphite-supported activated carbon electrodes. Right: typical pore size distribution of the activated carbon used.

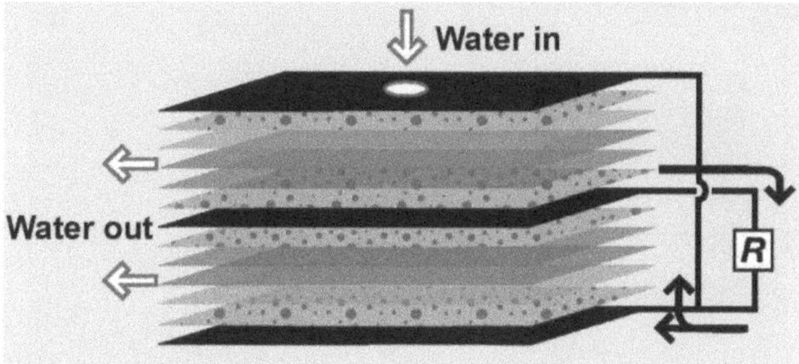

FIG. 7.9 Stack of electrode/membrane couples for CDP. *Reproduced with permission from Sales BB, et al. Direct power production from a water salinity difference in a membrane-modified supercapacitor flow cell. Environ Sci Technol 2010;44:5861. Copyright 2010, American Chemical Society.*

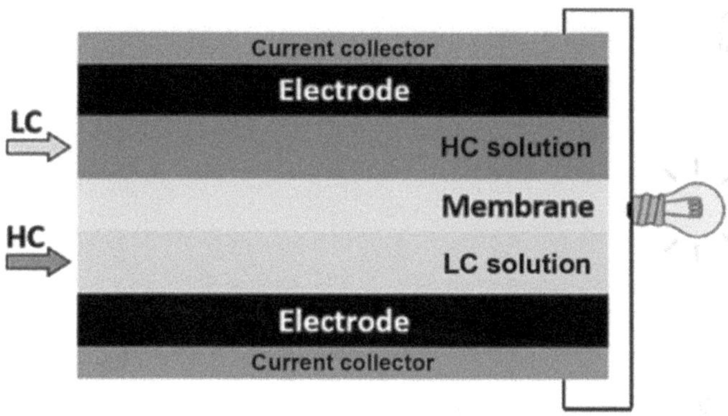

FIG. 7.10 The concentration capacitor based on two electrodes in contact with high- and low-salinity solutions separated by a membrane (it can be a filtration or an anion-exchange one). The solutions are forced to flow alternatively along each of the channels, for performing of the CapMix cycles. *Reproduced with permission from Zhan F, et al. High performance concentration capacitors with graphene hydrogel electrodes for harvesting salinity gradient energy. J Mater Chem A 2018;6:4981. Copyright 2018, American Chemical Society.*

on the current collector. The block and its microscopic structure are pictured in Fig. 7.11. The authors claim that it is the high porosity and surface area of the GHs, evidenced by these pictures, that explains the superior performance of this cell.

The principles of energy harvesting when a simple filtration membrane is used for separating the two streams are simple: as shown in Fig. 7.10, both sodium and chloride ions will flow from the upper channel by diffusion to the lower one; at the same time, an upward stream of sodium is also produced to fill the upper electrode and empty the lower one. As a result, a net flow of chloride ions in the downward direction will occur, leading to the formation of negative charge in the bottom electrode, and vice versa in the upper one.

FIG. 7.11 Pictures of the graphene hydrogel and its microscopic structure. *Reproduced with permission from Zhan F, et al. High performance concentration capacitors with graphene hydrogel electrodes for harvesting salinity gradient energy. J Mater Chem A 2018;6:4981. Copyright 2018, American Chemical Society.*

To this, one must add the junction potential of the membrane originated by the different diffusivities of Na^+ and Cl^-. With these reasonings, it can easily be understood that the effect will be enhanced by exchanging the filtration membrane for an ion exchange one. In such a case, the authors find a power production of almost 500 mW/m^2, well above the results obtained by other CapMix methods.

The experimental implementation of the SE technique is also easy, in principle [8]. The activated carbon particles are first deposited on the conducting substrate together with the adhesive polymer and the electrodes thus prepared are maintained during 24 h in contact with 0.02 M (on a monomer basis) solutions of the respective polyelectrolytes. In our case, we used the anionic poly(sodium 4-styrenesulfonate) (PSS) with molecular weight 70,000, and the cationic poly(diallyl-dimethylammonium chloride) (PDADMAC) (Fig. 7.12). Other polyelectrolytes were tested in [9], but the results were not significantly different. After rinsing with water, the electrodes were placed in the measuring and the cycle was started after a few hours short-circuit in salt water.

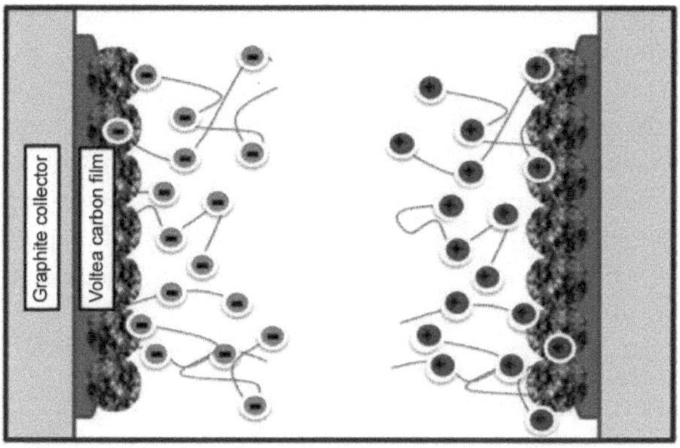

FIG. 7.12 A schematic picture of the SE configuration. *Reproduced with permission from Fernández MM, et al. Polyelectrolyte- versus membrane-coated electrodes for energy production by Capmix salinity exchange methods. J Power Sources 2016;302:387. Copyright 2016, Elsevier B.V.*

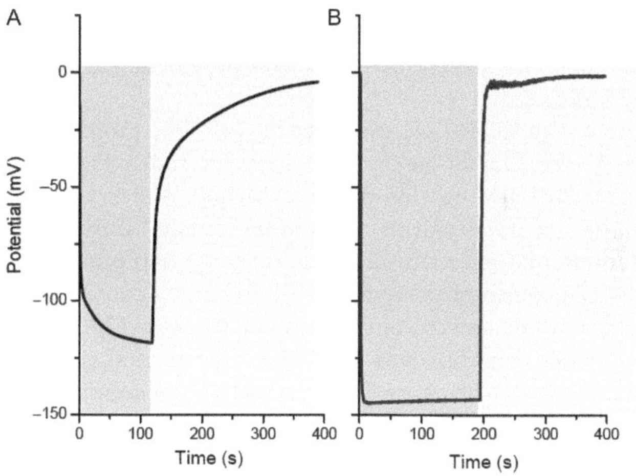

FIG. 7.13 Open circuit measurements for soft (A) and membrane-coated (B) electrodes when we alternately introduce salt and fresh water in the working cell. Dark-shaded areas: 30 g/L NaCl; light-shaded: 1 g/L. *Reproduced with permission from Fernández MM, et al. Polyelectrolyte- versus membrane-coated electrodes for energy production by Capmix salinity exchange methods. J Power Sources 2016;302:387. Copyright 2016, Elsevier B.V.*

A comparison between the results obtained with CDP and SE was reported in [9], and Fig. 7.13 shows that the two methods are similar regarding the voltage oscillation amplitude (around 120 mV for SE and 140 mV for CDP), but differ clearly regarding the kinetics: the former needs close to 200 s for voltage stabilization, while CDP requires only a few seconds. This clearly points to similar values of energy per cycle (20 mJ for CDP and 15 mJ for

SE), but better performance of CDP regarding power density (105 mW/m^2 vs. 50 mW/m^2). The reasons for such differences seem to rely on the way in which the two polyelectrolytes coat the activated carbon. EDX determinations demonstrated that, while PDADMAC coats the carbon film rather uniformly, PSS does so in patches where it penetrates the carbon. This indicates that it is this polymer that can be made responsible for the slower response. More research is needed on this issue, as the advantage of SE regarding the cost and ease of preparation associated with not needing membranes can make such additional investigation worthwhile.

7.6 Improvement: Wire Electrodes

As mentioned, although not as significant as in CDLE, the ohmic losses in CDP and SE are also a factor that plays against the performance of the method, common in fact to other electrochemical techniques and devices. For instance, as pointed out by Andreas [21], most of the energy of the supercapacitor is delivered in the high half of the nominal voltage, so even if the discharge is only partial, the energy available is close to residual. Hence a number of attempts have been proposed to minimize this unwanted effect, including coating the electrodes with polymers, surfactants, or even liquid crystals [22, 23].

In the context of CDP [24], Sales et al. [25, 26] proposed for the first time the use of wire-shaped instead of planar electrodes. Fig. 7.14 illustrates the principles of this new version of the technique. A slurry of activated carbon in a solution of polyvynilidene fluoride/1-methyl 2-pyrrolidone is first cast on a titanium/platinum wire 2 mm in diameter and 6 cm long and dried. The ion exchange membranes are then built in situ by contacting the carbon-coated wires in anionic and cationic ionomers, respectively. Contact with the sea and river water solutions was performed by immersing a pair of wires, externally connected with a resistor, in flasks containing the solutions (Fig. 7.14). It was demonstrated in [24, 26] that the power increased as the separation between wires decreased, ranging from 70 μW/g for 19 mm surface-to-surface separation to almost 350 μW/g for 1 mm separations. This improvement was ascribed to the important reduction in internal resistance upon wire approach.

Burheim et al. [25] modelled the process and demonstrated that mass transport by diffusion is enhanced in wire geometry as compared to plate-like electrodes, to the extent that pumping is not necessary and its corresponding power is largely saved, increasing the overall efficiency. In the absence of convection, it is diffusion that controls the salt concentration on the membrane surface, c_{mem}. Assuming that the cell has been short-circuited in salt water and so maintained for a sufficient time, the Donnan potential on each wire-membrane composite will be given by the logarithm of the ratio between c_{mem} and c_{sea}. The key point is the evaluation of c_{mem} as a function of time

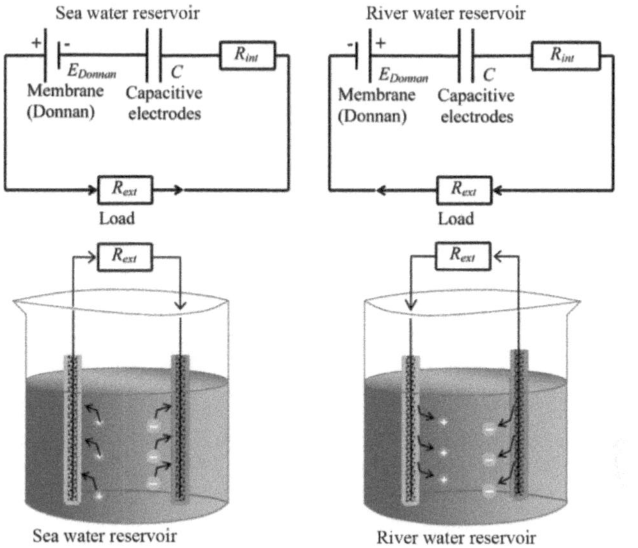

FIG. 7.14 Schematics of the wire electrode geometry for CDP. *Reproduced with permission from Sales BB, et al. Impact of wire geometry in energy extraction from salinity differences using capacitive technology. Environ Sci Technol 2012;46:12203. Copyright 2012, American Chemical Society.*

when the salt solution is abruptly changed from c_{sea} to c_{river} or vice versa. The model assumes that cylindrical symmetry is applicable, and that the second Fick's law can then be written:

$$\frac{\partial c_{salt}}{\partial t} = D_{salt}\left(\frac{\partial^2 c_{salt}}{\partial r^2} + \frac{1}{r}\frac{\partial c_{salt}}{\partial r}\right) \tag{7.13}$$

where D_{salt} is the salt diffusion coefficient in water (1.6×10^{-9} m^2/s for NaCl at room temperature). An important and realistic hypothesis is that right after the exchange of solutions a film of the previously contacted water remains adhered to the wire. The thickness of that slab, δ, relative to the radius of the cylinder, determines the time needed for the establishment of the steady-state open-circuit voltage (OCV). The value best fitting the time evolution of the OCV was found to be $\delta = 111$ μm. Finally, the fact that the steady-state OCV is reached in about 13 s for river and 3.7 s for sea water requires the additional hypothesis that the thickness mentioned should decrease when fresh water is substituted by salt water, and reciprocally.

7.7 Increasing the Power: Additional Charging

We have repeatedly mentioned earlier in this chapter that the main advantages of using membrane-coated electrodes relate to both a reduction of self-discharge and the possibility of a self-maintained cell voltage, controlled

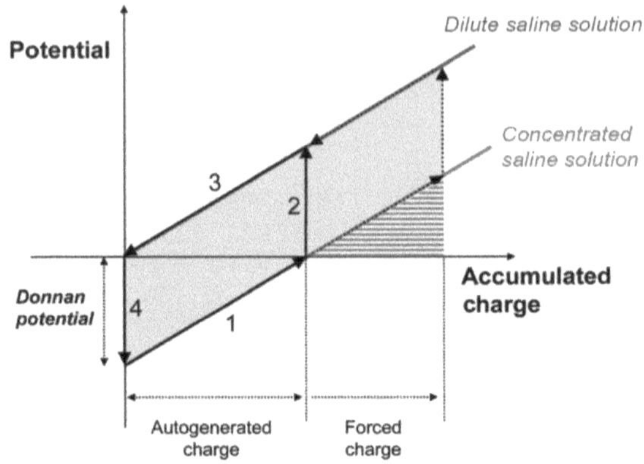

FIG. 7.15 Schematics of the potential versus transferred charge in a CDP process. The areas marked indicate: energy produced (shadowed) and energy spent from the external source (stripped). *Reproduced with permission from Liu F, et al. Effect of additional charging and current density on the performance of capacitive energy extraction based on Donnan potential. Energy Env Sci 2012;5:8642. Copyright 2012, Royal Society of Chemistry.*

by the Donnan potential on both sides of the membrane. Here we consider the possibility of increasing the power available by using an external power source enlarging the area of the voltage-charge cycles even at the price of the power consumed by the source. This was first proposed by Liu et al. [27], and is schematically shown in Fig. 7.15. A simple model was elaborated by these authors, leading to the following expression for the power produced:

$$P = \frac{W_{cycle}}{2t + t_{r,s} + t_{s,r}} \tag{7.14}$$

where W_{cycle} is the area under the whole V-Q experimental cycle, t is the charging or discharging time, and $t_{r,\ s}(t_{s,\ r})$ is what the authors call *switching time*, or time taken for the exchange between river and sea (or sea and river) solutions. The energy per cycle can in turn be expressed in terms of the integrals of the cell potential with respect to charge in the charging (negative contribution, it costs us energy, superscript C) and discharging (positive, D) steps:

$$W_{cycle} = -\int_0^Q \Delta V_{cell}^C dQ + \int_Q^0 \Delta V_{cell}^D dQ \tag{7.15}$$

and the cell potential in each step depends on the membrane potential $\Delta\Psi_{mem}^{C,D}$, the cell resistance $R_{cell}^{C,D}$ and capacitance C, and the intensity applied I:

$$\Delta V_{cell}^{C,D} = 2\Delta\Psi_m^{C,D} + \Delta V^{rise,drop} \pm IR_{\text{int}}^{s,r} - \Delta V_{leakage} \tag{7.16}$$

We assume here that we are charging in sea water and discharging in river water. It is important to consider, as demonstrated in [27], that the voltage (and energy) leakage increases with the amount of transferred charge, and

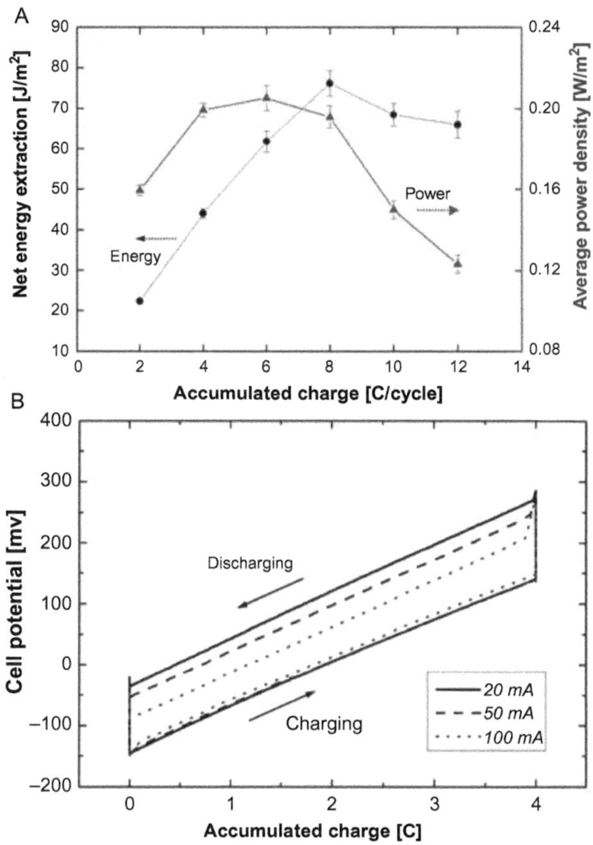

FIG. 7.16 (A) Energy and power densities in a forced charge CDP experiment with 50 mA applied current, and different values of the transferred charge. (B) Voltage versus charge if 4 C are transferred at the indicated intensities. *Reproduced with permission from Liu F, et al. Effect of additional charging and current density on the performance of capacitive energy extraction based on Donnan potential. Energy Env Sci 2012;5:8642. Copyright 2012, Royal Society of Chemistry.*

hence with the current applied for that charge transfer. As a result, it is not worth increasing the current, or the charging time, or the total charge, beyond a certain limit, beyond which the energy and power densities start to decrease. This feature is shown in Fig. 7.16A: the leakage increases with charge to the extent of producing a negative effect on the net energy, which hence decreases above a certain charge (roughly 8 C in the case shown). Additionally, Fig. 7.16B shows that the area of the cycle decreases when the current is increased from 20 to 100 mA, a clear consequence of the $IR_{\text{int}}^{s,r}$ term in Eq. (7.16): considering that the internal resistance is almost 10 times higher in river water, the negative effect of the resistance in river water is mainly observed, with little effect in the charging step in sea water.

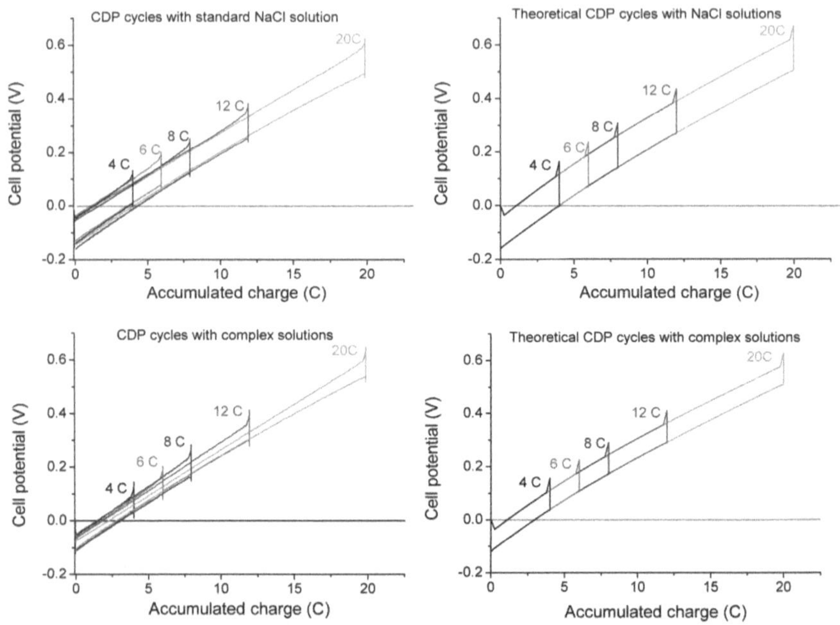

FIG. 7.17 Experimental data (left panels) and theoretical predictions (right panels) of CDP cycles for different amounts of externally injected charge. Charging in 511.7 mM and discharging at 551.7/30 mM of NaCl (top) and artificial sea water (bottom). *Reproduced with permission from Ahualli S, et al. Effect of solution composition on the energy production by capacitive mixing in membrane-electrode assembly. J Phys Chem C 2014;118:15590. Copyright 2014, American Chemical Society.*

A microscopic analysis of the mass and charge transfer through the membrane and electrode, along the lines of the one presented in Section 7.3 of this chapter, was elaborated by Ahualli et al. [5] for the case of forced charging. The results show a good quantitative agreement with experimental data, as demonstrated in the top panels of Fig. 7.17. In this work, the authors also considered the interesting practical situation that it is not NaCl solutions that are alternately pumped, but solutions with a composition mimicking that of typical sea waters, namely, 400 mM NaCl, 20 mM $MgSO_4$, 20 mM $MgCl_2$, 10 mM KCl, 10 mM $CaCl_2$, 1.7 mM KBr. Again, a reasonable agreement is found between data and model, as both indicate a reduction of the amount of energy (area enclosed by the cycles), a direct consequence of the 25% reduction of the cell potential when multivalent ions are present.

This feature is in fact explained in [5] on the basis of results like those depicted in Fig. 7.18. Both divalent cations, Ca^{2+} and Mg^{2+}, are transported *uphill*, or against the direction of diffusion implied by their corresponding gradients. This behavior is justified in different investigations of transport through membranes, based on the fact that the fastest-diffusing ions (Na^+ and K^+) create an electric field repelling the slow ones.

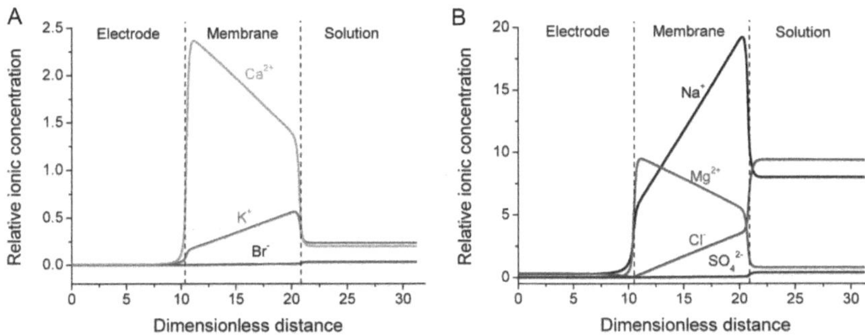

FIG. 7.18 Theoretical concentration profiles for the different ions in an artificial sea water in a cation exchange membrane with 1.5 M fixed charge. The solution concentration is 511.7 mM on the right of the membrane, and 1/30th of this value on the left. Note the different scales in (A) (less abundant ions) and (B). *Reproduced with permission from Ahualli S, et al. Effect of solution composition on the energy production by capacitive mixing in membrane-electrode assembly. J Phys Chem C 2014;118:15590. Copyright 2014, American Chemical Society.*

7.8 Increasing the Power: Stacking of Wire Electrodes

We have discussed earlier the interesting idea of using dipping wire-shaped electrodes instead of the usual plate-like ones. In this section of the chapter, the essential problem of increasing energy and power will be dealt with in the light of the ease of use of these electrodes, as demonstrated by Liu et al. [28]. The focus is again on the energy losses of the system. The authors demonstrate (Fig. 7.19) that nonohmic losses (self-discharge of the porous carbon electrode, membrane charge loss, etc.) are virtually independent of the number of associated electrode pairs, whereas ohmic ones (voltage drops in the internal resistance of the cell) are reduced in fresh water solutions, due to additional current paths between pairs of electrodes in close proximity.

7.9 Efficiency

Yip et al. [29] have carefully revised the efficiency issues regarding this and other CapMix techniques. A first estimate of efficiency can be reached by considering the maximum energy that could be theoretically obtained, which thermodynamically corresponds to the free energy of mixing, ΔG_{mix}, for the sea and river waters, with salt concentrations c_s and c_r, and corresponding volumes V_s and V_r. Considering that the latter volume is typically much larger than the former [30],

$$\Delta G_{mix} = -2RTV_r \left(c_s + c_r \ln \frac{c_r}{c_s} \right) \tag{7.17}$$

or about 2 MJ/m^3 of river water. The efficiency of the process, ε, could be estimated as the ratio between the energy obtained and the free energy reduction, relative to the same volume of solution. In the case of CDP, for the small cells

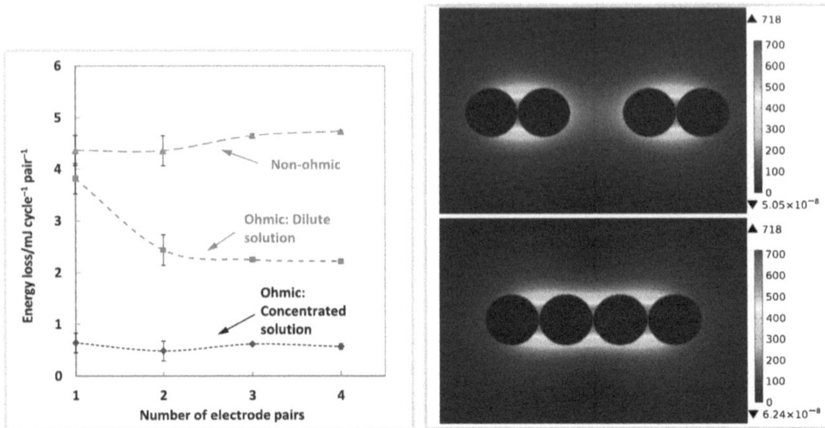

FIG. 7.19 Energy losses (both ohmic and nonohmic) in a parallel association of four wire-electrode pairs. The main reduction is of course associated with dilute (river) solution. Simulations indicate that current paths exist between electrodes when they are sufficiently close. *Reproduced with permission from Liu F, et al. Parallel up-scaling of capacitive mixing (CapMix) system enhances the specific performance. Electrochim Acta 2016;187:105. Copyright 2016, Elsevier Ltd.*

typically used, it is difficult to estimate ε, as the mixed volume in each cycle is not straightforward to determine. Nevertheless, Ahualli et al. [31] estimated that energy harvesting by double layer expansion is on the order of 5.4 J/L of fresh water, which represents about 1% of the thermodynamic maximum ($\varepsilon = 0.01$). In CDP, this value is about four times larger. It must be noted that, as mentioned in [29], a compromise must be found between increasing ε and maximizing the power: the former means working close to thermodynamic equilibrium, whereas the latter is achieved by reducing the cycle duration, thus increasingly further from quasi-equilibrium conditions. Analyses of the thermodynamic efficiency have been carried out for reverse electrodialysis [32–34] and pressure-retarded osmosis [35–39]. In the former technique, it was predicted that the efficiency can be as high as 37% for optimum selection of parameters. In the latter, a thermodynamic analysis [38] shows that no more than 75% of the ideal free energy of mixing can be achieved. In addition, pretreatment of the feed solutions, pumping for circulating them, and, eventually, biofouling of the membranes can make the technique impractical for large-scale operation.

Alternatively, a measure of efficiency can be proposed considering a theoretical model of the voltage-charge cycle using the calculated voltage rise associated with the salinity exchange, based on the charge balance measured. The area of the experimental cycle is in fact compared to the one expected in absence of losses, and the efficiency then rises to 30% [31]. It appears difficult to improve on this figure by CDP methods.

It is practical to express the energy or power relative to the apparent (macroscopic) area of the electrodes or other active elements. This allows comparison between the methods. Thus, while pressure retarded osmosis can reach up to 5 W/m^2, and reverse electrodialysis has the target in about 3 W/m^2, CapMix methods are now an order of magnitude lower, that is, around 200 mW/m^2. Economic implications and feasibility are beyond the scope of the present study, and can be found in [29].

In the case of using soft electrodes, the results in [8] can shed some light as to the efficiency figures. Fig. 7.20 depicts the actual results compared to theoretical ones. The power obtained was 12.1 ± 0.1 mW/m^2 (relative to the apparent electrode area), or an energy per cycle of 0.58 mJ. This corresponds to a surface charge density sweep (Fig. 7.20C) of 0.3 μC/cm^2. From the current-time data in Fig. 7.20B, the total charge transferred per cycle is 18 mC, this indicating that the actual area of carbon participating in the process is around 6 m^2, a minor portion of the micropore area (roughly 200 m^2 for 0.2 g of carbon on the electrodes). This is one of the reasons why the energy produced by CapMix methods is lower than theoretically available: only a small fraction of interfacial area is in reality at work [40].

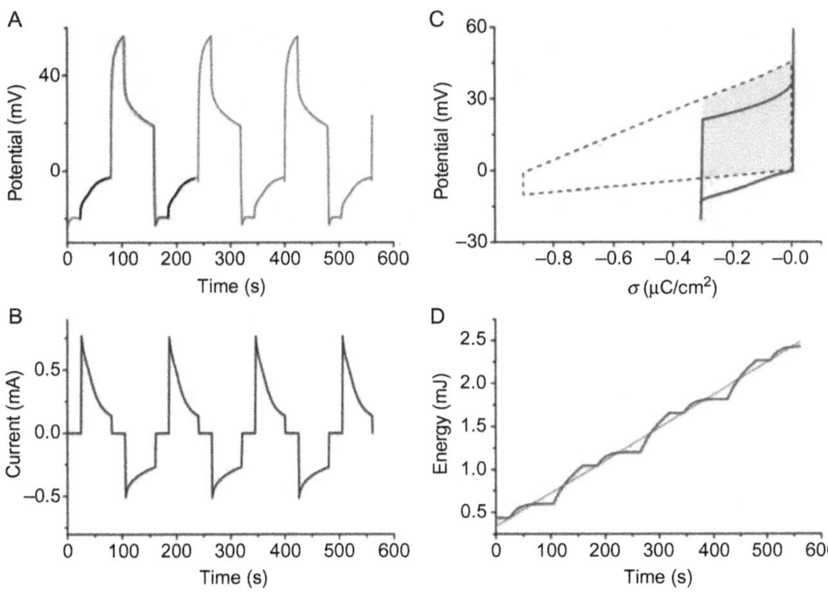

FIG. 7.20 (A) Voltage and (B) current oscillations when salt and fresh solutions are exchanged in the SE cell; (C) the potential vs. surface charge density measured (solid line), and calculated (dashed line). The shaded representation corresponds to the portion of the theoretical cycle actually covered by experimental data; (D) total energy accumulated in successive cycles as a function of time. *Reproduced with permission from Ahualli S, et al. Polyelectrolyte-coated carbons used in the generation of blue energy from salinity differences. Phys Chem Chem Phys 2014;26:25241. Reproduced by permission of the PCCP Owner Societies.*

The efficiency can again be calculated by comparing the power to that theoretically available from the solution mixing rate:

$$\varepsilon = \frac{\langle VI \rangle}{\Delta G_{mix} \Phi_v} \tag{7.18}$$

where $\langle VI \rangle$ is the average power, Φ_v is the pumping flow rate (1 mL/s), and ΔG_{mix} is the mixing free energy per unit volume. The result is $\varepsilon = 0.5\%$, comparable to values previously reported for CDP. As mentioned, it is possible to obtain a practical efficiency by comparing the area of the experimental cycle and that theoretically available, as performed in Fig. 7.20C: the result is now 36%.

7.10 Conclusions

In this chapter, we have revised the existing approaches to obtaining blue energy from electrical double layers using membranes or polyelectrolyte coatings. The advantage over CDLE of not requiring external power sources is common to all the techniques described, and the amount of energy and power obtained are among the highest obtained by purely electrical double layer based methods. Nevertheless, it is clear that the energy obtained is far (1% or so) from that thermodynamically available, and it is apparent that only a small fraction of the area of the micropores is actually working in the production of energy. The selection of materials appears to be a fundamental aspect, as well as the implementation at larger scale. In this case, energy costs of pumping and, eventually, water pretreatment must be considered in establishing a realistic balance of the future of this kind of energy.

References

[1] Conway BE. Electrochemical supercapacitors. Scientific fundamentals and technological applications. New York: Plenum Press; 1999.

[2] Conway BE, Pell WG, Liu TC. Diagnostic analyses for mechanisms of self-discharge of electrochemical capacitors and batteries. J Power Sources 1997;65(1-2):53–9.

[3] Sales BB, Saakes M, Post JW, Buisman CJN, Biesheuvel PM, Hamelers HVM. Direct power production from a water salinity difference in a membrane-modified supercapacitor flow cell. Environ Sci Technol 2010;44(14):5661–5.

[4] Biesheuvel PM, Wal A. Membrane capacitive deionization. J Membr Sci 2010;346(2):256–62.

[5] Ahualli S, Fernandez MM, Iglesias G, Jimenez ML, Liu F, Wagterveld M, Delgado AV. Effect of solution composition on the energy production by capacitive mixing in membrane-electrode assembly. J Phys Chem C 2014;118(29):15590–9.

[6] Makino K, Ohshima H, Kondo T. Surface-potential of an ion-penetrable charged membrane. J Theor Biol 1987;125(3):367–8.

[7] Ohshima H, Kondo T. Membrane-potential and Donnan potential. Biophys Chem 1988; 29(3):277–81.

[8] Ahualli S, Jimenez ML, Fernandez MM, Iglesias G, Brogioli D, Delgado AV. Polyelectrolyte-coated carbons used in the generation of blue energy from salinity differences. Phys Chem Chem Phys 2014;16(46):25241–6.
[9] Fernandez MM, Wagterveld RM, Ahualli S, Liu F, Delgado AV, Hamelers HVM. Polyelectrolyte-versus membrane-coated electrodes for energy production by Capmix salinity exchange methods. J Power Sources 2016;302:387–93.
[10] Duval JFL, Ohshima H. Electrophoresis of diffuse soft particles. Langmuir 2006;22(8):3533–46.
[11] Dukhin SS, Zimmermann R, Werner C. Electrophoresis of soft particles at high electrolyte concentrations: an interpretation by the Henry theory. J Colloid Interface Sci 2007;313(2):676–9.
[12] Ohshima H. Electrokinetics of soft particles. Colloid Polym Sci 2007;285(13):1411–21.
[13] Ohshima H. Donnan potential and surface potential of a spherical soft particle in an electrolyte solution. J Colloid Interface Sci 2008;323(1):92–7.
[14] Ohshima H. The Donnan potential-surface potential relationship for a cylindrical soft particle in an electrolyte solution. J Colloid Interface Sci 2008;323(2):313–6.
[15] Ohshima H. Theory of electrostatics and electrokinetics of soft particles. Sci Technol Adv Mater 2009;10(6):063001.
[16] Ahualli S, Luisa Jimenez M, Carrique F, Delgado AV. AC electrokinetics of concentrated suspensions of soft particles. Langmuir 2009;25(4):1986–97.
[17] Burheim O, Sales BB, Schaetzle O, Liu F, Hamelers HVM. Auto generative capacitive mixing for power conversion of sea and river water by the use of membranes. J Energ Resour-Asme 2013;135(1):483–92.
[18] Zhan F, Wang ZJ, Wu TT, Dong Q, Zhao CT, Wang G, Qiu JS. High performance concentration capacitors with graphene hydrogel electrodes for harvesting salinity gradient energy. J Mater Chem A 2018;6(12):4981–7.
[19] Xu YX, Sheng KX, Li C, Shi GQ. Self-assembled graphene hydrogel via a one-step hydrothermal process. ACS Nano 2010;4(7):4324–30.
[20] Lim HN, Huang NM, Lim SS, Harrison I, Chia CH. Fabrication and characterization of graphene hydrogel via hydrothermal approach as a scaffold for preliminary study of cell growth. Int J Nanomedicine 2011;6:1817–23.
[21] Andreas HA. Self-discharge in electrochemical capacitors: a perspective article. J Electrochem Soc 2015;162(5):A5047–53.
[22] Xia M, Nie J, Zhang Z, Lu X, Wang ZL. Suppressing self-discharge of supercapacitors via electrorheological effect of liquid crystals. Nano Energy 2018;47:43–50.
[23] Tevi T, Yaghoubi H, Wang J, Takshi A. Application of poly (p-phenylene oxide) as blocking layer to reduce self-discharge in supercapacitors. J Power Sources 2013;241:589–96.
[24] Liu F, Advancing CapMix for electricity generation. Ph.D. thesis, Wageningen University, Wageningen; 2015.
[25] Burheim OS, Liu F, Sales BB, Schaetzle O, Buisman CJN, Hamelers HVM. Faster time response by the use of wire electrodes in capacitive salinity gradient energy systems. J Phys Chem C 2012;116(36):19203–10.
[26] Sales BB, Burheim OS, Liu F, Schaetzle O, Buisman CJN, Hamelers HVM. Impact of wire geometry in energy extraction from salinity differences using capacitive technology. Environ Sci Technol 2012;46(21):12203–8.
[27] Liu F, Schaetzle O, Sales BB, Saakes M, Buisman CJN, Hamelers HVM. Effect of additional charging and current density on the performance of Capacitive energy extraction based on Donnan Potential. Energy Environ Sci 2012;5(9):8642–50.

[28] Liu F, Donkers TFW, Wagterveld RM, Schaetzle O, Saakes M, Buisman CJN, Hamelers HVM. Parallel up-scaling of capacitive mixing (CapMix) system enhances the specific performance. Electrochim Acta 2016;187:104–12.

[29] Yip NY, Brogioli D, Hamelers HVM, Nijmeijer K. Salinity gradients for sustainable energy: primer, progress, and prospects. Environ Sci Technol 2016;50(22):12072–94.

[30] Veerman J, Reverse electrodialysis. Design and optimization by modeling adn experimentation. Ph.D. thesis, University of Gröningen, Gröningen, The Netherlands; 2010.

[31] Ahualli S, Fernandez MM, Iglesias G, Delgado AV, Jimenez ML. Temperature effects on energy production by salinity exchange. Environ Sci Technol 2014;48(20):12378–85.

[32] Logan BE, Elimelech M. Membrane-based processes for sustainable power generation using water. Nature 2012;488(7411):313–9.

[33] Vermaas DA, Veerman J, Yip NY, Elimelech M, Saakes M, Nijmeijer K. High efficiency in energy generation from salinity gradients with reverse electrodialysis. ACS Sustain Chem Eng 2013;1(10):1295–302.

[34] Yip NY, Vermaas DA, Nijmeijer K, Elimelech M. Thermodynamic, energy efficiency, and power density analysis of reverse electrodialysis power generation with natural salinity gradients. Environ Sci Technol 2014;48(9):4925–36.

[35] Yip NY, Elimelech M. Performance limiting effects in power generation from salinity gradients by pressure retarded osmosis. Environ Sci Technol 2011;45(23):10273–82.

[36] Yip NY, Elimelech M. Thermodynamic and energy efficiency analysis of power generation from natural salinity gradients by pressure retarded osmosis. Environ Sci Technol 2012;46(9):5230–9.

[37] Kim YC, Elimelech M. Potential of osmotic power generation by pressure retarded osmosis using seawater as feed solution: analysis and experiments. J Membr Sci 2013;429:330–7.

[38] Lin SH, Straub AP, Elimelech M. Thermodynamic limits of extractable energy by pressure retarded osmosis. Energy Environ Sci 2014;7(8):2706–14.

[39] Bar-Zeev E, Perreault F, Straub AP, Elimelech M. Impaired performance of pressure-retarded osmosis due to irreversible biofouling. Environ Sci Technol 2015;49(21):13050–8.

[40] Guillermo MMF, Iglesias R, Ahualli S, Jiménez ML, Kozynchenko OP, Delgado ÁV. Materials selection for optimum energy production by double layer expansion methods. J Power Sources 2014;261:371–7.

Section E

Capacitive Deionization (CDI)

Chapter 8

Principles and Theoretical Models of CDI: Experimental Approaches

Silvia Ahualli, Guillermo R. Iglesias and Ángel V. Delgado
Department of Applied Physics, University of Granada, Granada, Spain

8.1 Introduction

The responsible use of water, as well as its recycling and purification, has been a major issue for our society for decades. Considering that the reserves of fresh water on the planet represent only 3% of the total available, obtaining water suitable for subsequent consumption is a more than attractive idea that has been studied recently.

The main process used, on the basis of growth and installed capacity worldwide, is reverse osmosis (RO), which primarily expanded thanks to the work of Loeb and Sourirajan [1]. Their work transformed membrane separation from a laboratory to an industrial process. The method is based on pumping salt solutions with high pressure through a semipermeable membrane, leaving behind between 95% and 99% of dissolved salts (in addition to other pollutants such as bacteria and pyrogens) and allowing the flow of clean water [2]. The higher the salt concentration, the larger the pressure that must be exerted to overcome the osmotic pressure. Other methods are multistage flash distillation, electrodialysis, vacuum distillation, or solar evaporation. The selection of the procedure involves estimating the energy needed to desalinate a given solution to acceptable levels. But all of these processes have high energy costs, either for mechanical application as in the case of RO or in heating the solution.

In 1960, a new concept of desalination was proposed for the first time by Blair et al. [3], called *electrochemical demineralization of water*. This work was the starting point of what is known as capacitive deionization (CDI is the usual acronym), nowadays admitted to be an efficient and low-cost alternative for desalination of brackish water (up to 10 g/L NaCl). It is based on the high electrical capacitance of the electric double layer (EDL; see Chapter 1) and the consequent possibility of adsorbing comparatively large amounts of

Interface Science and Technology, Vol. 24. https://doi.org/10.1016/B978-0-12-811370-7.00008-5

ions in solution. This is somehow the key to its low energy consumption: ions are extracted from water instead of extracting water from the concentrated salt solution. This has brought about scientific interest in the method during the last decade, which is reflected in the number and quality of scientific articles in specialized journals of high impact in the area.

Anderson et al. [4] have made theoretical calculations regarding the work required to produce 1 m^3 of a 300 mg/L NaCl solution from solutions of different concentrations: CDI can be economically and energetically feasible for desalinizing solutions with concentrations below 5000 mg/L for efficiencies ranging in the moderate range of 60%–70%. Nevertheless, this figure needs to grow up to 85% if CDI has to actually compete with RO in deionizing solutions with salinity comparable to seawater.

In addition, CDI can be used not only for desalination of water for human or agricultural uses [5]; for instance the method has been proposed for the separation of insulin from zinc chloride ($ZnCl_2$) [6], which is one of the impurities associated with the purification process of insulin, traditionally separated by chromatographic methods. Another study has shown advantages in using CDI for water and chemical savings to desalinate the feed water stream in cooling towers [7]. In this chapter we provide an overview of the fundamentals and main experimental implementations of this deionization technique.

8.2 Basic Principles

Imagine a pair of electrodes immersed in an electrolyte solution, subjected to an externally applied potential difference. The charge that is stored as excess electrons in one of the electrodes and defect of electrons in the other, attracts counterions until it is compensated for by adsorption of ions of opposite sign coming from the solution that bathes both electrodes. This process is highly reversible, and hence the benefits of the method. The existence of the electronic charge at the solid surface and the ionic charge in the adjacent solution constitutes the EDL. This structure is well known, although in the case of the porous electrodes to be used it presents particular aspects, as the fact that a notorious ion adsorption is only achieved in the so-called micropores, which are those with diameters on the order of 2 nm that must necessarily constitute each electrode to obtain the enormous capacitances that are achieved [8]. Fig. 8.1 represents the ionic order that can be expected in these pores, as well as the distribution of electrical potential. On the other hand, we can distinguish another type of pores, the *macropores*, denoting the interparticle space between carbon particles, serving as pathways for ion transport across the electrode (Chapters 1 and 2).

The ions adsorbed on the respective EDLs are removed from the solution in the interelectrode channel, leaving it partially deionized. The key point is the capacity of the EDL to adsorb ions, hence the denomination given to the method. The accumulation of ions continues until the electronic charge of the electrodes is completely screened, so a crucial aspect will be the achievement

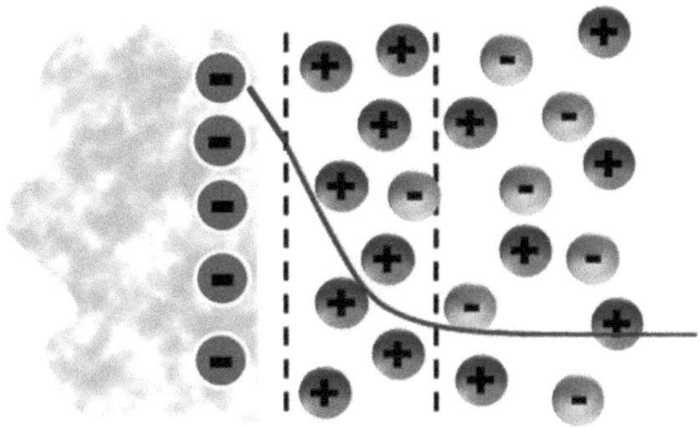

FIG. 8.1 Ionic distribution and electric potential at the surface of a micropore.

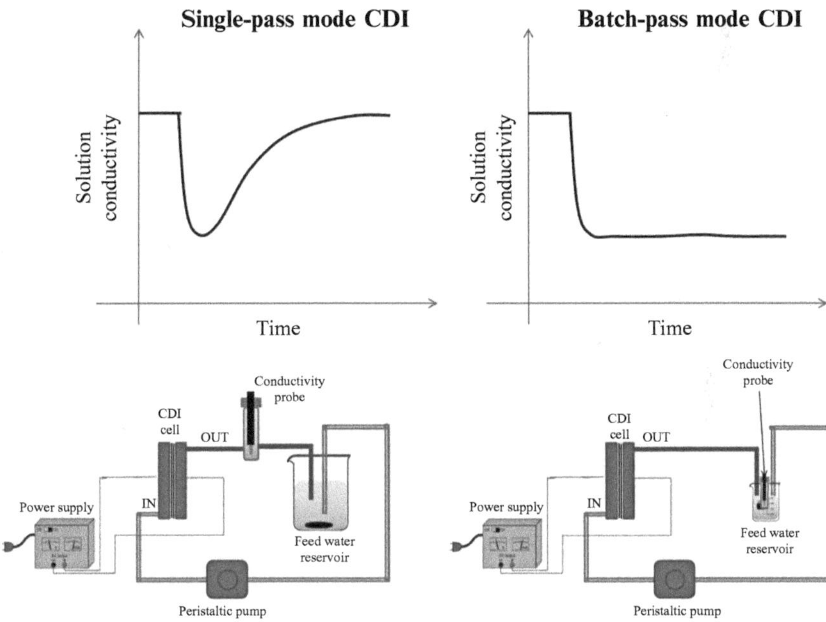

FIG. 8.2 Schematic diagram of single-pass (left) and batch-pass (right) modes for CDI experiments.

of a very large exposed surface, capable of adsorbing as many ions as possible and removing them from the solution. Note that the counterions of each electrode are attracted to it, while coions are released from it. However, the total content of salt in solution decreases, since coions are in lower concentration and are also adsorbed in their counter-electrode. The result is that it is possible to reach admissible salt concentrations in drinking water (0.5–1 g/L Nacl).

The basis of the process is represented in Fig. 8.2. During the desalination cycles, the ionic solution is pumped from the container into the cell, while at its output continuous measurements of the conductivity and temperature are performed. In the case of mixed solutions, the outgoing solution is sampled at specified time intervals, and the ionic composition of the samples is determined by ion chromatography. There are different ways to carry out the experiment and they can be divided according to different criteria:

- According to the experimental layout: The circulation of the solution through the desalination cell can be performed in two different ways, namely, *single-pass mode* or *batch-pass mode* (Fig. 8.2). In the former, the conductivity is measured in line at the output of the cell. This quantity drops soon when ions migrate to the electrodes in the adsorption step. The outgoing solution can be discarded or recycled back to the same reservoir if this is large enough in such a way that its conductivity does not change. When the electrodes have reached their adsorption capacity, the effluent salinity rises again, approaching the feed conductivity. In the second case (batch-pass mode), the feed water reservoir is much smaller, enough to appreciate changes in conductivity, which is measured in that same container. During the adsorption step, the conductivity decreases until it reaches a minimum when the pores become saturated. The initial value is not recovered as in single-pass mode, because the reservoir volume is small and at the end it gets a constant, low conductivity upon being partially desalinated.
- According to the power source used we distinguish between *constant current* (CC) or *constant voltag*e (CV) (Fig. 8.3). If a constant voltage is applied, at the initial time the current flowing through the circuit is maximum and ions begin to move towards the electrode of opposite sign. The current subsequently decreases, while the micropores reach a maximum storage value, and the charge is screened. Gradually, the system returns to a steady state (zero current, constant conductivity) because the electrodes cannot adsorb further ions. In Fig. 8.3 a representation of the cycle can be observed. In the desorption step, the current goes in the opposite direction and the micropores start to be emptied, releasing the ions into the channel, which in turn reaches a concentration greater than the incoming one. Once the micropores are emptied, the concentration in the channel tends to the value of the feed solution. When operating under constant-current mode a constant ionic conductivity is achieved (salinity composition) at the output of the cell. However, a constant level of salt composition can only be obtained under CC operation if performing the experiment in combination with ion exchange membranes (Membrane CDI, to be described shortly) and not in CDI alone [8].
- In order to return the ions to the solution and leave the micropores empty, the electrodes are externally short-circuited (*zero voltage*, *ZV*, or *zero current*, *ZC*) (Fig. 8.4). Another option is the application of a voltage difference that is inverted with respect to the previous step (*reverse voltage*,

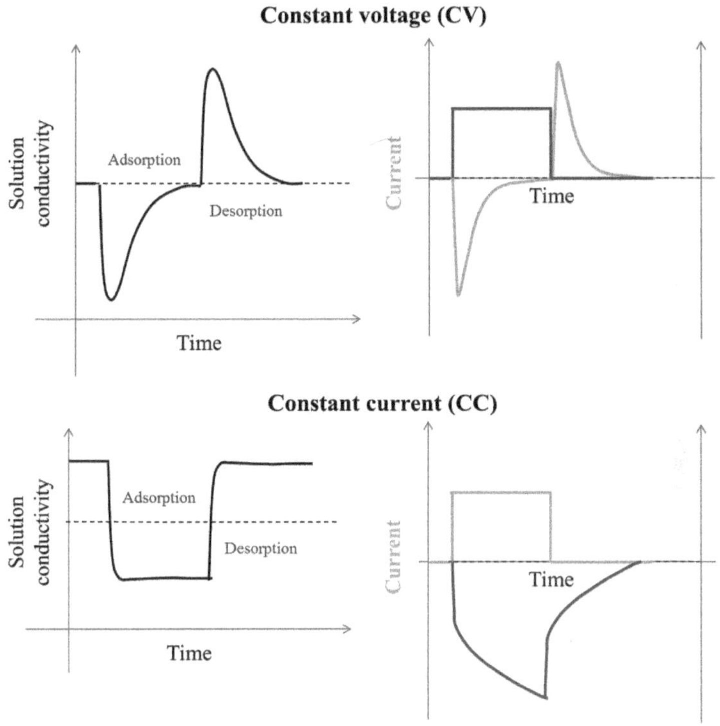

FIG. 8.3 CDI cycle for constant voltage (CV), and constant current (CC) operation. Input voltage or current and output signals (current or voltage and ionic conductivity) are represented.

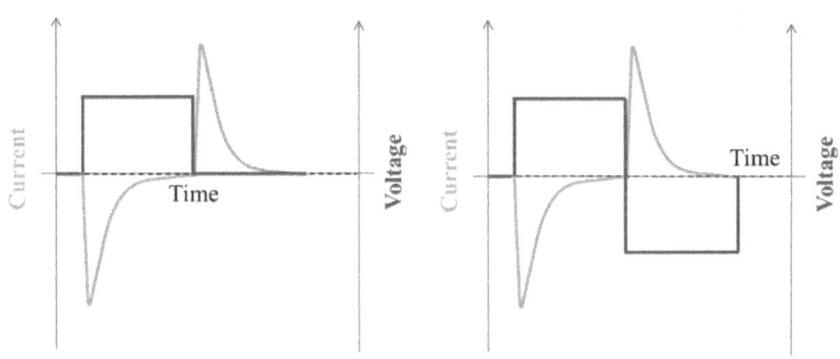

FIG. 8.4 Current and voltage during adsorption and desorption steps in zero-voltage mode (ZV) (left) and reverse-voltage mode (RV) (right).

RV, or *reverse current*, *RC*), in such a way that the adsorbed ions become coions of their electrode, and are repelled by the electronic charge thereof. The electric force will cause a migration of the ions towards the opposite electrode. This requires considerable differences regarding the cell architecture used, as we will see later.

An alternative to the classical solution flowing between two parallel electrodes is a cell in which the solution flows *through* the electrodes, in the same direction as the applied electric field [9, 10]. This arrangement of the cell, known as *flow-through electrode* or *flow-through CDI*, has the advantage of not needing a spacer where the solution flows. Instead, an electrode/separator/electrode assembly replaces the classical configuration, thus decreasing the internal resistance of the cell, and making a faster desalination possible as a consequence of the reduction in the time needed for salt removal from the electrodes [11].

One of the determining factors in the CDI process is the voltage applied to the electrodes. Fig. 8.5 displays a comparison of the concentration differences found for two different applied voltages. On the one hand, it is observed that by increasing the voltage difference between the electrodes, the concentration of the outlet solution reaches lower values, that is, the solution loses more ions. The area under the curve, related as will be seen later to the adsorbed salt, is greater for higher voltages. By applying a larger potential difference, the charge on the electrodes increases and a larger number of ions will be necessary to compensate for this charge. As we know from the theory of EDL, the charge–voltage relationship is not linear, hence higher voltages do not guarantee better results (some authors report up to 1.4 V), and furthermore faradaic reactions will take place.

Another key factor is the ionic concentration of the feed solution. It is expected that by increasing it, adsorption does not change appreciably, since it depends on the volume of available micropores [12]. In fact, rather than the specific surface area of the electrode, the ion adsorption is determined by the ionic size and specific volume [13], in other words, the maximum capacitance of the EDL. Hence, the use of the CDI technique is limited to moderately salty solutions, and it is impractical to consider it for desalination

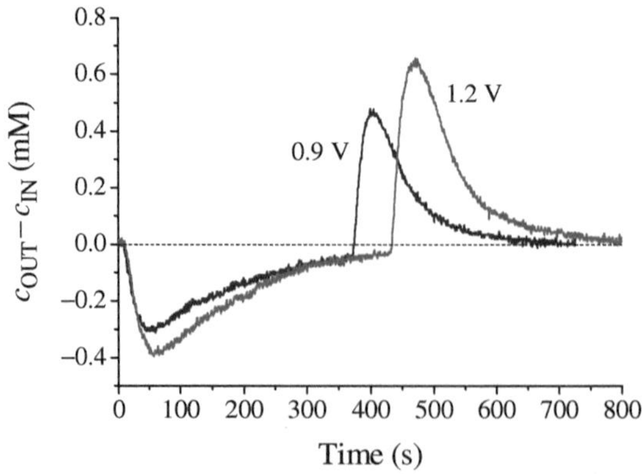

FIG. 8.5 Ionic concentration difference between the outlet and inlet of the cell by applying 0.9 V and 1.2 V.

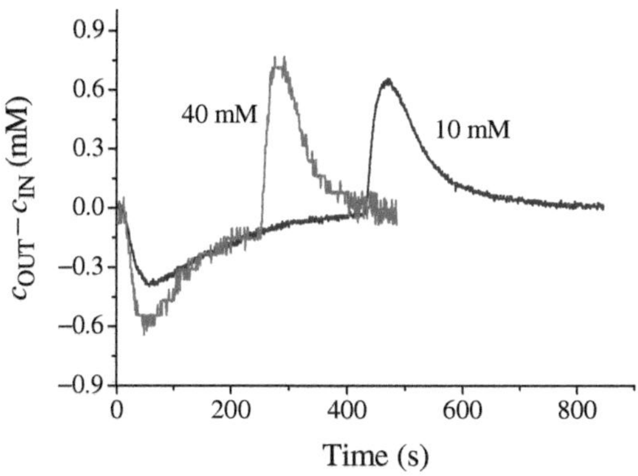

FIG. 8.6 CDI cycle for ionic concentration feed solutions of 10mM and 40mM.

of seawater. A back-of-the-envelope estimation can be performed as follows: reducing the ionic strength from 500mMNaCl to 20mMNaCl means removing about 20g of salt from 1L of solution; for typical experimental values of the salt adsorption capacity (20mg of salt per gram of carbon, see the following text), this means about 1kg of carbon in the electrodes.

However, due to the overlapping of the EDLs inside micropores, a nonlinear increase with feed solution ionic concentration is found (Fig. 8.6). Note that the decrease of the cell resistance at higher salt concentration brings about a faster kinetics and thus a shorter cycle [14–16].

8.3 CDI Performance Evaluation

In order to quantify the parameters of the process, the following indicators have been defined [17]:

Salt adsorption capacity: The total number of adsorbed salt moles, n:

$$n = \int_0^{t_{ads}} (c_{out}(t) - c_{in})\phi_v dt \tag{8.1}$$

where ϕ_v is the pumping velocity of solution in the CDI circuit, and t_{ads} is the duration of the adsorption step. It is clear that, in the curve of concentration differences as a function of time, the area under the curve will represent the number of moles adsorbed in the cycle, n. It is thus possible to define the salt adsorption capacity, SAC, as the mass in grams of adsorbed ions per mass of electrodes:

$$SAC = n\frac{M_{salt}}{m_{elec}} \tag{8.2}$$

where M_{salt} is the molecular weight of salt and m_{elec} is the mass of both dried electrodes, including the additive binder.

Charge efficiency: The charging efficiency of the system, ε, is the ratio between the adsorbed salt and the transferred charge between electrodes:

$$\varepsilon = \frac{F\int_0^{t_{ads}} (c_{out}(t) - c_{in})\phi_v dt}{\int_0^{t_{ads}} I(t)dt} \tag{8.3}$$

where F is the Faraday constant. The maximum possible efficiency is 1, which would occur when one ion is removed for each unit of charge deposited on the electrodes [18]. Experimentally, the maximum is reached by increasing cell voltage or Stern capacitance, or by decreasing the ionic strength of the solution being treated.

Energy: The energy per ion, E, is the energy consumed during the desalination process per removed ion:

$$E = \frac{\int_0^{t_{ads}} VI(t)dt}{2\int_0^{t_{ads}} (c_{out}(t) - c_{in})\phi_v dt} \tag{8.4}$$

where V is the applied voltage in the CV mode.

8.4 Membrane Capacitive Deionization

8.4.1 Basic principles

In recent years, different types of cell architectures have been proposed, each with different characteristics pursuing a common target, namely optimizing the efficiency of the process. One of the most significant modifications is the use of ion exchange membranes, a method that receives the name of membrance capacitive deionization (MCDI), and that was first demonstrated by Lee et al. [19]. The solution flows between two electrodes, each one in contact with a selective membrane for cations (CEM) or anions (AEM) (Fig. 8.7) depending on the polarity of the electrode, whether negative or positive, respectively. In order to understand the role of membranes in CDI, it would be good to point out that the EDL is formed by increasing the concentration of counterions in the micropores at the same time that coions decrease in number in such a way that the electronic charge deposited at the interface is compensated. If there is an ion exchange membrane (IEM) between the solution and the electrode, most of the counterions will cross it and end up in the EDL, while the coions are

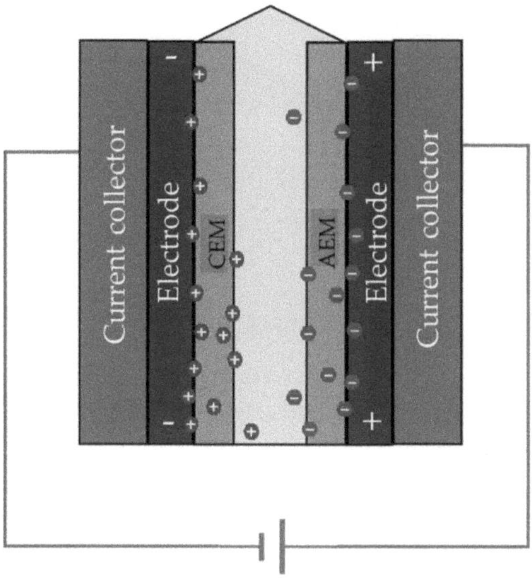

FIG. 8.7 Top: the ionic solution flows between cathode and anode close to cation-exchange (CEM) and anion-exchange (AEM) membranes. Bottom: schematics of ion migrations without (left) and with (right) ion exchangers.

blocked by it and cannot leave the electrode region. This causes two direct effects on the CDI technique. On the one hand, the requirement of electroneutrality brings as a consequence the need for a larger number of counterions to compensate for the electronic charge, and thus a greater number of ions are removed from the solution and, with them, the dissolved salt. On the other hand, the efficiency will increase because for each electron that is transferred between the electrodes, an ion will be removed from the solution (it will be part of the EDL of one of the electrodes), while without the membranes, not only the counterions are adsorbed but the coions can be expelled, ending up in the solution to be desalinated and diminishing the efficiency. Therefore, the origin of the improvement of the method lies in the blocking of the coions of the membrane itself.

As an example of our experimental results, we show in Fig. 8.8 the adsorption stage using carbon electrodes alone or in combination with IEMs. The use of membranes produces a greater adsorption (deeper peak) and a larger area, which is a direct measure of the adsorption, as can be seen from Eq. (8.2), and which is magnified with the applied voltage. An appreciable qualitative difference in the cycles with CDI and MCDI is obtained if the desorption cycle is carried out with reverse voltage (RV) (Fig. 8.9).

8.4.2 Internal structure

An IEM consists of a polymer matrix in which ionic groups are chemically bound to the polymeric backbone, making a coherent ion-exchanger gel [20].

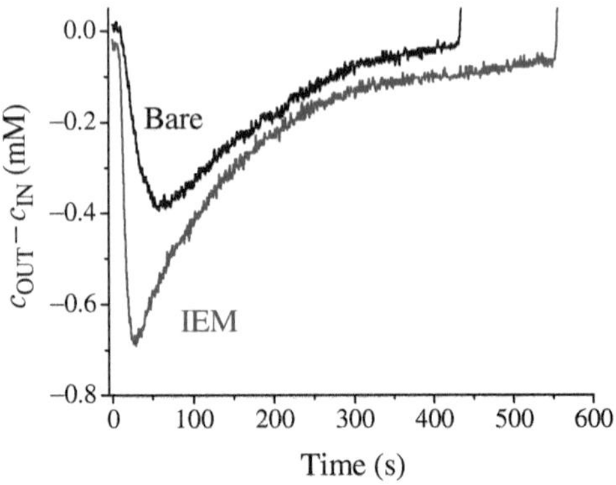

FIG. 8.8 Ionic concentration difference between the output and inlet of the cell in CDI (bare) and MCDI (IEM) configuration in ZV mode (1.2 V).

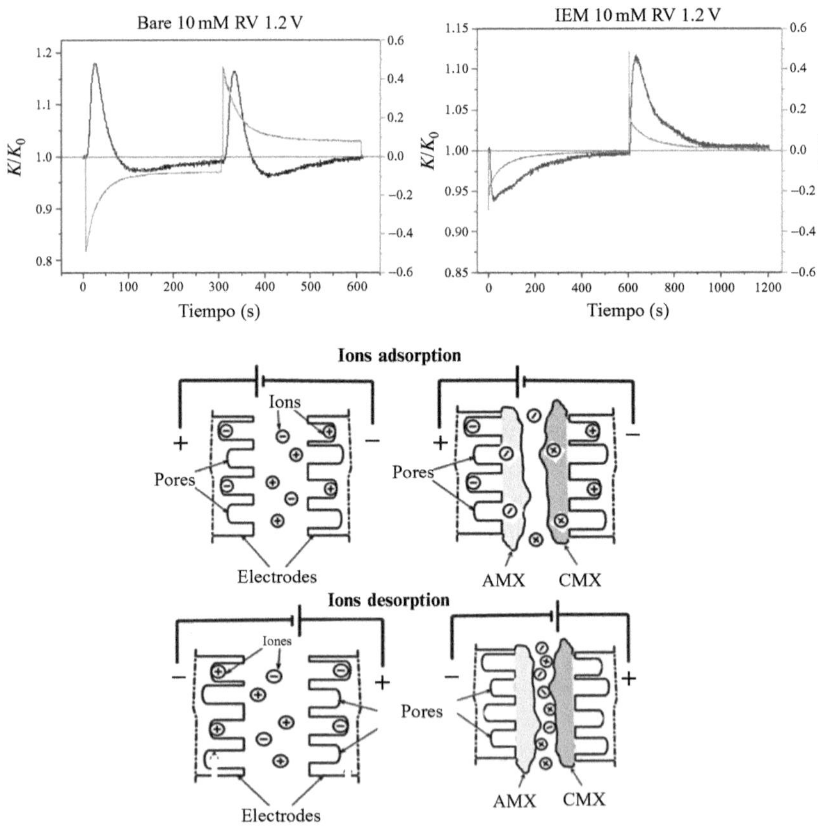

FIG. 8.9 Top: relative conductivity of 10 mM NaCl using reversed-voltage mode (RV) for CDI (left) and MCDI (right) in one cycle. Bottom: sketches of both situations. AMX (CMX) anion (cation) exchangers.

The charge of the ionic groups determines whether the membrane is either anion-exchange or cation-exchange. The positively charged groups fixed in the former can be NH_3^+, $R\text{-}NH_2^+$, $R_2\text{-}NH^+$, and $R_3\text{-}N^+$, which permit the transport of anions and reject that of cations. Concerning the negatively charged membranes (CEM), they typically contain such groups as SO_3^-, COO^-, PO_3^{2-}, and PO_3H^-, clearly favoring the transport of cations and rejecting anions. The increase in structural rigidity and of the density of fixed charged groups is attained by adding a crosslinker, typically divinyl benzene monomer.

The simple model of an IEM described in Refs. [21–23] and Chapter 7 allows the evaluation of the number of ions that can be withdrawn from an ingoing solution. We recall here that the membrane can be modeled as a flat region, highly charged due to the dissociation of fixed molecular groups, and characterized by its thickness, its volume charge, and its selective permeability. These parameters will determine the membrane potential and the ion flux established when ionic solutions with different concentrations bathe both sides of it. In order to calculate the charge and potential distributions in the ion-exchange membranes, we solve the Poisson equation governing the electric potential Ψ, in the region of the membrane, assuming that it contains a constant charge density, ρ_{memb}.

In Fig. 8.10 the predominance of counterions is observed, so that their larger concentration as compared to that of coions compensates for the charge of the membrane and determines the negligible flow of coions. The effect of ionic concentration of the solution determines the behavior of the membranes in MCDI: when the concentration of the feed solution increases, the flow of the coions ceases to be negligible, although it is still lower than that of the counterions. This will affect the efficiency of the MCDI cycles, that is expected to fall.

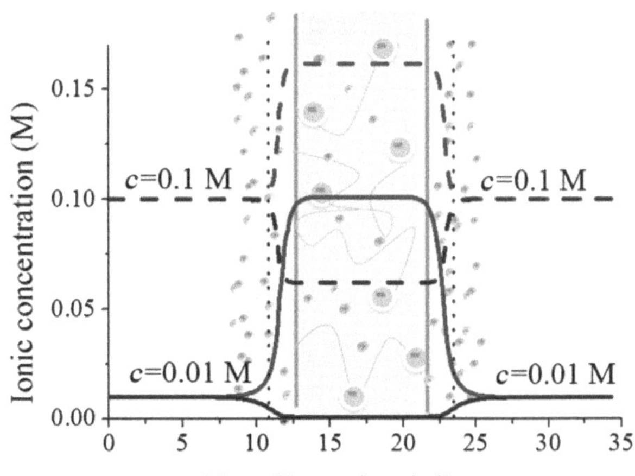

FIG. 8.10 Counterions (+) and coions (-) concentration profile for a charged membrane (0.1 M). The membrane is in contact with solutions of 0.01 M (solid lines) or 0.1 M (dashed lines).

8.4.3 Practical issues

From the previous information, it will be clear that the IEMs used in MCDI must have high permselectivity (a value of 1 means that no coions can go through the membrane), low resistance, and high stability, both chemical and mechanical. Several investigators have developed IEMs specifically for MCDI: Kwak et al. [24] produced a cation exchange membrane through heat crosslinking and esterification of three monomers: 4-styrenesulfonic acid sodium salt hydrate (NaSS), methacrylic acid (MAA), and methyl methacrylate (MMA). The synthesis of a poly(vinylidene fluoride)-sodium 4-vinylbenzene sulfonate copolymer (PVDF-g-PSVBS) CEM for deionization applications was also reported by Kang et al. [25]. Jeong et al. [26] synthesized an aminated poly(vinylidene fluride-g-4-vinylbenzyl chloride) (PVDF-g-VBC) anion exchange membrane, while Qiu et al. [27] used irradiation to crosslink a thin sulfonated polystyrene film as a CEM. Nevertheless, most often it is commercial IEMs that are used in the field.

MCDI provides more salt adsorption capacity and a higher charge efficiency compared to CDI, due to the almost complete elimination of coion adsorption. The increase ranges from 2 for $50\,mgL^{-1}$ of feed solution [28] to 1.5 for $400\,mgL^{-1}$ or $750\,mgL^{-1}$ [29, 30]. Such efficiencies depend on the solution concentration as shown in Fig. 8.11, where different CDI cell architectures are compared [20]. It can be seen that the use of IEMs keeps t+he MCDI efficiency above 80%, whereas the CDI technique yields a value below 70%. However, it is true that for more concentrated solutions, in the order of

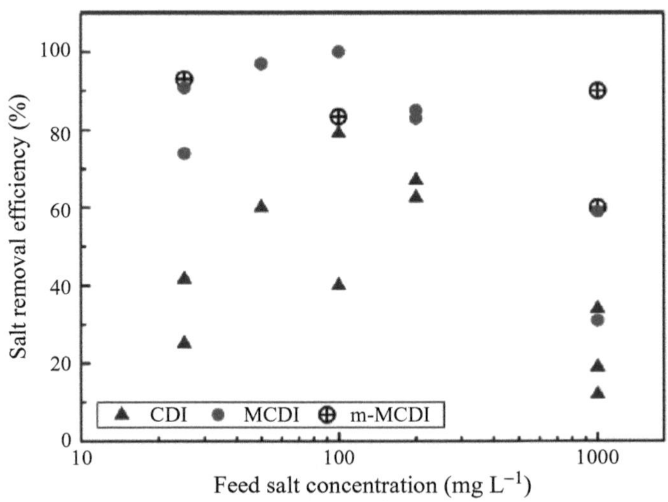

FIG. 8.11 Salt removal efficiency as a function of feed salt concentration using CDI, MCDI, and modified MCDI. *Reproduced with permission from Hassanvand A, et al.: The role of ion exchange membranes in membrane capacitive deionisation membranes, Membranes 7:54, 2017. Copyright 2017, MDPI.*

1000 mg/L, the efficiency falls and new architectures offer better results. This is the case of modified MCDI, to be described later.

It is worth considering the energy cost of CDI or MCDI operation. In both cases it will increase with the input solution concentration and decrease with efficiency. In any case, as shown in [29, 31], the power requirements are systematically higher in the case of simple CDI as compared to MCDI. In addition, Choi [32] demonstrated that the CC mode is preferable, in terms of energy consumption, to CV. Finally, compared to other desalination techniques like RO, Zhao et al. [33] found that MCDI is more energy efficient for dilute feed water concentration (less than 2 gL^{-1}) to obtain effluent water with concentration 0.5 gL^{-1} (Fig. 8.12).

8.4.4 Further applications of MCDI

The use of IEMs gives feasibility to the application of MCDI to selectively adsorb certain ionic species from water. Some studies on this matter are listed here:

- Sulfate ions are removed by single-pass RC, as sulfate is preferentially adsorbed compared to chloride [34]. The reason is found in the relation of the fluxes towards the electrode of SO_4^{2-} and Cl−. As shown in Chapter 7, the flux depends on the valence and diffusion coefficient in such a way that one type of ion is mainly transported to the anode with respect to the other. It has also been shown that, in general, MCDI is more efficient in withdrawing divalent than monovalent ions [35].

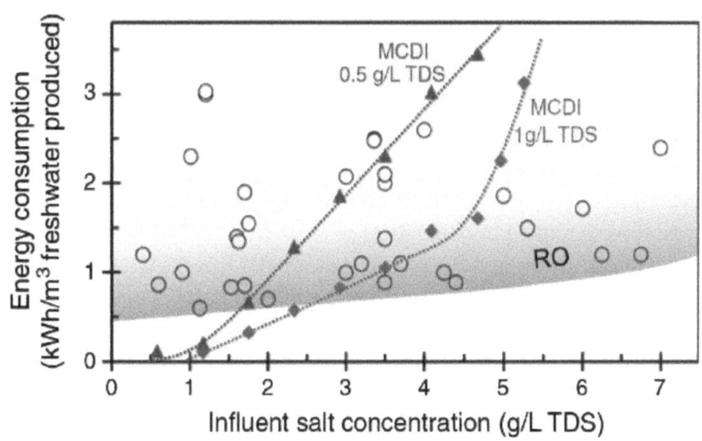

FIG. 8.12 Energy consumption as a function of feed water concentration for 0.5 gL^{-1} (triangles) and 1 gL^{-1} (diamonds) of total dissolved solids (TDS) using MCDI. Circles: data from RO technique. *Reproduced with permission from Zhao YJ, et al.: Energy consumption in membrane capacitive deionisation for different water recoveries and flow rates, and comparison with reverse osmosis, Desalination 330:35–41, 2013. Copyright 2013, Elsevier Ltd.*

- The combination of MCDI with a monovalent anion permselective exchange membrane has been applied to the removal of fluoride or nitrite from water [36, 37]. The experimental data show that using CDI the adsorption rate of F− is 9.7% and SO_4^{2-} reaches 12.6%. If membranes are superimposed (MCDI), the removal rates are 15.8% and 24.3%, respectively, revealing a preference for divalent ions. In addition, in the case of adding a monovalent anion permselective exchange membrane to prevent the transport of divalent ions, F− is adsorbed by 20.2% and SO_4^{2-} by 17.4%.
- Also, other types of ions such as nitrate [38, 39], phosphate [40], ammonium, calcium, and magnesium can be selectively extracted.

A practical problem of capacitive deionization is the fouling of either (or both) the electrodes or the IEMs. This undesirable phenomenon is caused by deposition of substances onto the electrode assembly, when they are capable of migrating towards them as the dissolved ions do. These substances can be colloidal particle clays, organic materials, or even bacteria [41]. Although this problem is also present in RO, according to Hassanvand et al. [20] it is more serious in CDI than in MCDI, as carbon electrodes are damaged in the former. As a rule, fouling will bring about a decrease in efficiency and an increase in power requirements [42].

8.5 Treated Electrodes

As discussed elsewhere in this text, finding new materials is essential for improving capacitive methods for either energy harvesting or solution deionization. The basic material has traditionally been activated carbon, or, more recently, other carbonaceous materials such as nanotubes or graphene. Lee et al. [43] were the first authors who combined an ion exchange resin with a CDI electrode with the aim of improving the overall hydrophilicity. However, later investigations have focused on coating the electrode with an ion selective layer [44], this resulting in selectivities comparable to the use of IEM and electrodes just adjacent. The approach, known as modified MCDI (m-MCDI), has energy and economical costs lower than conventional MCDI, even with comparable current efficiency. Other examples of coatings can be cited. Thus, Kim et al. [45] used a cross-linked poly(vinyl alcohol) (PVA) with sulfur succinic acid (SSA) and poly(styrene sulfonic acid-co-maleic acid) (PSSA_MA) (PVA/PSSA_MA/SSA) coating as cation exchange polymer and aminated polysulfone as anion exchanger, and used the resulting structure for the separation of NaCl, $CaSO_4$, and $MgCl_2$. In [46] Kim and Choi fabricated a coated electrode using cross-linked PVA- sulfosuccinic acid, and they showed that the charge efficiency of salt removal was around 75%–85%, higher than in CDI, where the figure was up to 67%. The improvement was attributed to the selectivity in cation transport, thanks to the cation exchange layer.

Another method resorting to treated electrodes has been denominated *inverted capacitive* deionization (i-CDI) [47]. In this method, the anode is modified by providing a negative surface charge to the anode through chemical treatment, while the cathode remains as pristine carbon material. The system behaves in an inverted fashion, that is, ions are released when the cell is externally charged, and vice versa. The method turned out to be significantly more efficient and stable than CDI.

In all of the studies mentioned, the use of IEMs superimposed on the carbon electrodes was replaced by their coating with polyelectrolyte. Inspired by the layer-by-layer deposition technique, another approach was proposed in [48, 49]: a layer of cationic and anionic polyelectrolyte onto each electrode is deposited without chemical manipulations [50, 51], converting the electrodes into a sort of "soft" electrode pair (Fig. 8.13), following the terminology used for colloidal particles lined by a polymer layer ("soft particles").

The process of electrode preparation is easy to perform. A graphite collector with a thin layer of activated carbon particles bound by poly(vinylidene-fluoride) (PVDF) adhesive is immersed in a 100-mM solution of a positive polyelectrolyte (PDADMAC) or a negative one (PSS) during 24 h under constant stirring. The resulting polyelectrolyte layer plays a similar role to the ionic exchange membranes, favoring the flux of counterions and hindering that of coions, thus improving the salt adsorption (Fig. 8.14), while keeping the efficiency of cycles close to 1, up to 1 V cell potential. That is why Fig. 8.14 shows a similar behavior to that achieved in the presence of IEMs in reverse mode (Fig. 8.9), confirming the blocking of coions in the same way as in MCDI. Furthermore, it has been shown by comparison of the current peaks during the cycles [50] that the polyelectrolyte layer does not introduce an extra electrical resistance in the system, in contrast to the case of membranes.

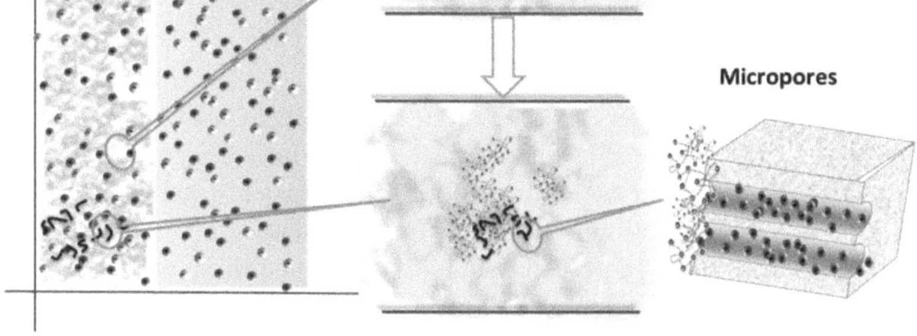

FIG. 8.13 Sketch of solution in contact with a "soft" electrode.

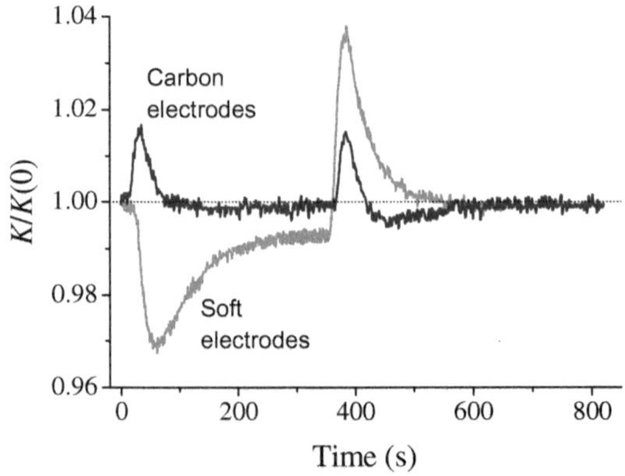

FIG. 8.14 Cycle comparing the relative conductivity of NaCl 20 mM using carbon electrodes with soft and bare electrodes in reverse-voltage mode (0.9 V).

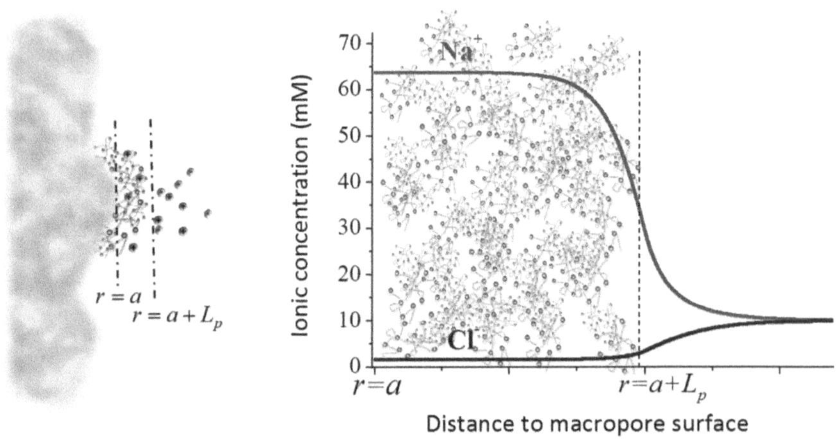

FIG. 8.15 Left: Sketch of macropore surface–polyelectrolyte layer–ionic solution interface. The region between $r = a$ y $r = a + L_p$ is occupied by the polymer. Right: Theoretical predictions of sodium (counterion) and chloride (coion) ionic concentration as a function of distance to the electrode surface. Ionic concentration of electrolyte solution: 10 mM. *Reproduced with permission from Ahualli S, et al.: Use of soft electrodes in capacitive deionization of solutions, Environ Sci Technol 51(9):5326–5333, 2017. Copyright 2017, American Chemical Society.*

In addition, the polyelectrolyte coating gives rise to a hydrophilization of the electrode surface, and brings about further benefits such as the "protective" effect of the carbon surface by minimizing electrochemical reactions. As pointed out before [47, 52], repetition of the CDI cycles with the very same pair of electrodes demonstrates large reproducibility and stability, with consistently greater salt removal than polymer-free electrodes.

The polyelectrolyte layer can be patterned similarly to a membrane (Fig. 8.15): a region of space with a volumetric charge density where ions can penetrate. However, this layer is deposited on the electrode surface itself, so that the electrolytic solution bathes the layer only on one side thereof. As before, the potential distribution and the ionic profiles can be obtained by solving the Poisson equation in the region outside the solid interface (Chapter 7).

A similar behavior to that of the ion exchange membrane is observed (Fig. 8.15): the concentration of coions notoriously decreases while there is an excess of counterions in the region occupied by the polyelectrolyte. This means that the polymer layer will affect the transport of ions from the macropore to the surface of the electrodes, promoting the transport of counterions and hindering that of the coions.

8.6 Flow Electrodes

Recall that both CDI and MCDI require two steps in each cycle: adsorption and desorption. It is impossible to perform the adsorption if the micropores have not been emptied in a previous step. The performance of the CDI technique depends to a large extent on how efficient this emptying is. As an example, by applying reverse voltage instead of short-circuiting the electrodes, ions are forced to leave the electrode, where they become coions, and this provokes a rise of salt adsorption.

On the other hand, a challenge of the CDI method is that the process of solution desalination is cost-intensive because of the large micropore area needed, despite the use of IEMs or modified electrodes. However, a novel and promising cell architecture was proposed that might eliminate the salt from the feed water, with no need for a discharging step, and therefore it could be used to treat water with a high salt concentration. This technique is called flow-electrode capacitive deionization (FCDI), and was proposed for the first time in 2013 by Jeon et al. [53].

Briefly, two current collectors are connected to a power source in such a way that ions migrate through the ion-exchange membrane towards the suspended activated carbon particles (typically 10 μm in size), where they can be eventually adsorbed. Thus, in this new approach, traditional static electrodes are replaced by suspensions of micron-scale activated carbon particles in an electrolyte being pumped through electrode compartments. The discharge of the active carbon particles occurs downstream of the cell as a separate process (Fig. 8.16).

Note first that this is a milestone in capacitive deionization; recall that in standard CDI the concentration in the desalination stage reaches a minimum value and then rises, again reaching the initial value. In FCDI, in contrast, because the electrode material flows continuously through the cell, removing the adsorbed ions out of it, the concentration decreases, reaching a plateau. Furthermore, the carbon particles can be reused by mixing the two oppositely charged streams and letting the mixture settle out of the salt solution, while the brine is discarded. It is noteworthy that, even if the salt concentration is relatively high (Fig. 8.17), the FCDI method is able to perform satisfactorily. By way of example, it has been shown [53] that it is possible to desalinate NaCl up to 35 gL^{-1}, according to Fig. 8.17.

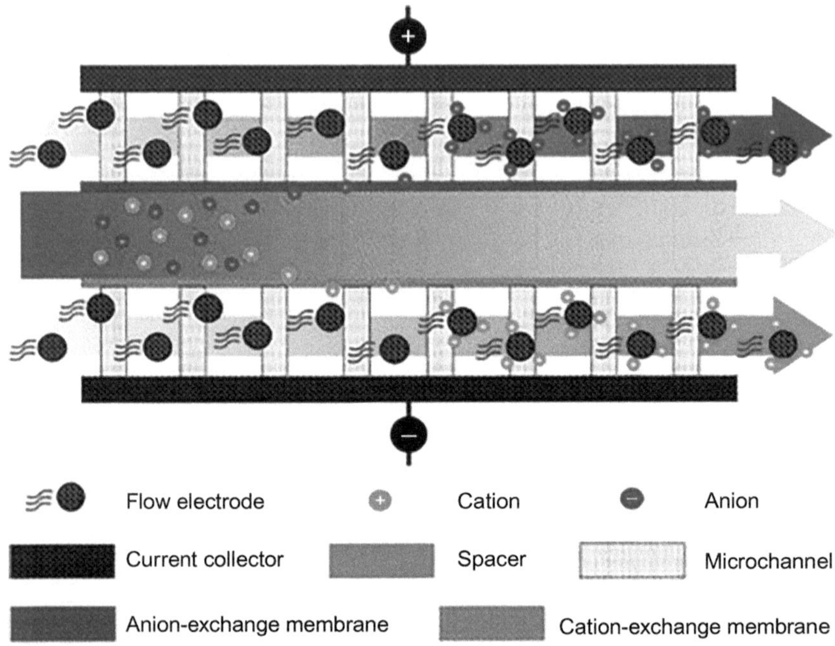

FIG. 8.16 Schematic representation of flow-electrode CDI process: The ionic solution flows between the electrodes arrangement. The ion exchange membranes separate the solution on one side from the carbon suspension on the other side. The carbon suspension flows in the microchannel capturing ions on the way and releasing them out of the cell in a separate container. *Reproduced with permission from Jeon SI, et al.: Desalination via a new membrane capacitive deionization process utilizing flow-electrodes. Energy Environ Sci 6:1471, 2013. Copyright 2013, Royal Society of Chemistry.*

An important limitation of the method is the low electronic conductivity of the carbon slurry, on the order of 1 mS/cm at the most [54], because of the discontinuous network of particles. This value is much lower than that of static carbon plug electrodes. This limits the transport of electronic charge and hence the rate of salt removal and the energy efficiency. This drawback can be partially compensated for by increasing the carbon concentration of the slurry. This would reduce both the electronic resistance (electron transfer from collectors to particles) and the ionic one (ions transfer from the solution to the particles). Improvement has indeed been observed for concentrations of about 20% [54]; this is about the upper limit, since the flowability of the carbon suspension is compromised for larger content of particles, due to the change in its rheological properties (viscosity increase, clogging) [54, 55].

Again, the surface treatment helps the capacitive deionization methods: Park et al. [56] synthesized activated carbons with functional groups on their surface, namely, trimethylammonium (AC-N) for the anode and sulfonate

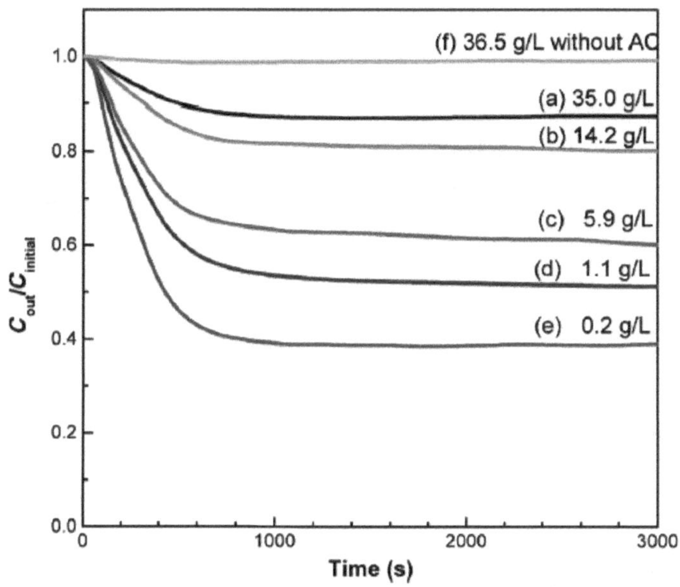

FIG. 8.17 Ratio between the outlet and initial concentration of NaCl for different feed solution concentration: 0.2, 1.1, 5.9, 14.2, and 35.0 g L^{-1}. The salt removal efficiencies were 61.0%, 48.8%, 39.9%, 19.9%, and 12.4%, respectively. Applied voltage: 1.2 V. *Reproduced with permission from Jeon SI, et al.: Desalination via a new membrane capacitive deionization process utilizing flow-electrodes, Energy Environ Sci 6:1471, 2013. Copyright 2013, Royal Society of Chemistry.*

(AC-S) for the cathode. This treatment of the carbon surface permits an increase in the concentration of the flow electrode, up to 35%, increasing in turn the salt removal efficiency to above 27%. The problem of aggregation and viscosity was minimized by the electrostatic repulsion between carbon particles in the same stream. On top of this, the coating favoured the dispersibility of the originally hydrophobic carbon particles.

A different procedure can still be mentioned: Doornbusch et al. [57] could reach 35% carbon concentration in the electrode compartment, while maintaining low concentrations (2.5%) in the surroundings, an architecture known as *Fluidized Bed Electrodes* (FbedCDI), which prevents clogging of the suspension and reduces pumping power requirements. The basis of this approach is the use of comparatively large (>100 μm diameter) spherical carbon beads in a flow cell disposed in a vertical orientation (Fig. 8.18) in such a way that the suspension is pumped against gravity. This leads to an accumulation of particles in the bottom of the cell, where they build up dense arrangements in the electrode compartments. The resulting fluidized bed adsorbs the ions, and the particles are afterwards driven into a mixing container where they release the adsorbed ions via collisions between oppositely charged beads.

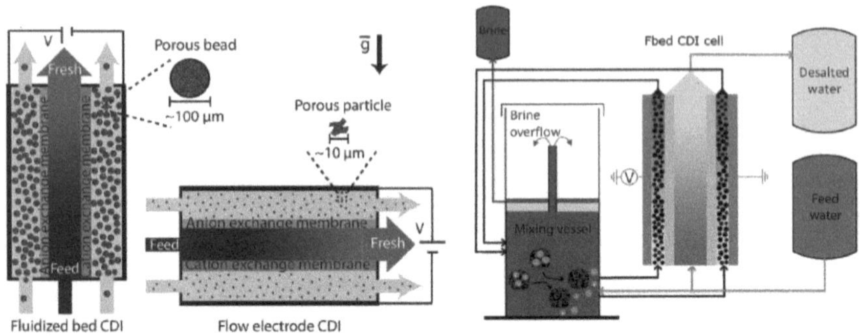

FIG. 8.18 Schematic differences between FCDI and FbedCDI (left) and (right) complete circuit of FbedCDI. *Reproduced with permission from Doornbusch GJ, et al.: Fluidized bed electrodes with high carbon loading for water desalination by capacitive deionization, J Mater Chem A 4:3642–3647, 2016. Copyright 2016, Royal Society of Chemistry.*

8.7 Conclusions

In this chapter we have revised the fundamental aspects of the desalination methods based on the large electrical capacitance of EDL, jointly known as capacitive deionization (CDI) methods. The general basis of these techniques is indeed the fact that a significant number of ions can be adsorbed on the solid/liquid interfaces if the area of such interfaces is high enough (like in porous electrodes) and a moderate potential difference is applied between opposite electrodes. The chapter has provided information about the different working strategies, including batch-pass vs. single-pass modes, constant voltage vs. constant current when charging, and zero-voltage vs. reversed-voltage during desorption. After reviewing the quantitative measures of salt withdraw efficiency (salt adsorption capacity and charge and energy efficiencies), the advantages of the use of ion exchange membranes (MCDI) resting on the carbon film have been discussed. A related technology has been denominated soft electrodes (SE), based on the production of a layer of polyelectrolyte (respectively, cationic or anionic, depending on the electrode) coating the activated carbon electrodes. Both MCDI and SE yield larger salt removal rate and efficiency thanks mainly to the fact that coions of the respective membrane or polyelectrolyte layer are not allowed to go through, and hence the cation and anion separation is largely improved. A particular advantage is that coions are blocked in reversed voltage mode, avoiding their adsorption on the opposite electrode and hence controlling the discharging step, essential to the method. Further improvements are related to flow- (or flowable) electrodes: the ion adsorption and release occurs on carbon particles flowing in the solution chambers separated by ion exchange membranes. The limitations associated with the rheological and stability drawbacks of very concentrated slurries have been partially eliminated by the use of a fluidized-bed electrode architecture, whereby the cell is kept vertical and the particles are allowed to settle on the bottom of the electrode chamber and are not transported along the channel. Jointly considered, the results certainly indicate that CDI-based methods are a reality and can be a good alternative to costly methods of water solution deionization in certain locations and communities where access to large-scale techniques is limited.

Acknowledgments

Financial support for this work from Junta de Andalucía, Spain (Project PE2012-FQM-0694), and RYC-2014-16901 (MINECO) is gratefully acknowledged. S.A. acknowledges the support from the University of Granada ("Vicerrectorado de Investigación, Plan Incorporación Jóvenes Doctores").

References

[1] Loeb S, Sourirajan S. Saline water conversion-II. Adv Chem Ser 1964;38:117.
[2] Shannon MA, Bohn PW, Elimelech M, Georgiadis JG, Marinas BJ, Mayes AM. Science and technology for water purification in the coming decades. Nature 2008;452(7185):301–10.
[3] Arnold BB, Murphy GW. Studies on electrochemistry of carbon and chemically modified carbon surfaces. J Phys Chem 1961;65(1):135–8.
[4] Anderson MA, Cudero AL, Palma J. Capacitive deionization as an electrochemical means of saving energy and delivering clean water. Comparison to present desalination practices: will it compete? Electrochim Acta 2010;55(12):3845–56.
[5] Oren Y. Capacitive delonization (CDI) for desalination and water treatment—past, present and future (a review). Desalination 2008;228(1-3):10–29.
[6] Jung SM, Choi JH, Kim JH. Application of capacitive deionization (CDI) technology to insulin purification process. Sep Purif Technol 2012;98:31–5.
[7] van Limpt B, van der Wal A. Water and chemical savings in cooling towers by using membrane capacitive deionization. Desalination 2014;342:148–55.
[8] Porada S, Zhao R, van der Wal A, Presser V, Biesheuvel PM. Review on the science and technology of water desalination by capacitive deionization. Prog Mater Sci 2013;58(8):1388–442.
[9] Avraham E, Bouhadana Y, Soffer A, Aurbach D. Limitation of charge efficiency in capacitive deionization. J Electrochem Soc 2009;156(6):P95–9.
[10] E. Avraham, M. Noked, Y. Bouhadana, A. Soffer, D. Aurbach, Limitations of charge efficiency in capacitive deionization II. On the behavior of CDI cells comprising two activated carbon electrodes, J Electrochem Soc 156(10) (2009) P157-P162.
[11] Suss ME, Biesheuvel PM, Baumann TF, Stadermann M, Santiago JG. In situ spatially and temporally resolved measurements of salt concentration between charging porous electrodes for desalination by capacitive deionization. Environ Sci Technol 2014;48(3):2008–15.
[12] Kim T, Dykstra JE, Porada S, van der Wal A, Yoon J, Biesheuvel PM. Enhanced charge efficiency and reduced energy use in capacitive deionization by increasing the discharge voltage. J Colloid Interface Sci 2015;446:317–26.
[13] Porada S, Borchardt L, Oschatz M, Bryjak M, Atchison JS, Keesman KJ, Kaskel S, Biesheuvel PM, Presser V. Direct prediction of the desalination performance of porous carbon electrodes for capacitive deionization. Energy Environ Sci 2013;6(12):3700–12.
[14] Yang KL, Ying TY, Yiacoumi S, Tsouris C, Vittoratos ES. Electrosorption of ions from aqueous solutions by carbon aerogel: An electrical double-layer model. Langmuir 2001;17(6): 1961–9.

[15] Ying TY, Yang KL, Yiacoumi S, Tsouris C. Electrosorption of ions from aqueous solutions by nanostructured carbon aerogel. J Colloid Interface Sci 2002;250(1):18–27.

[16] Noked M, Avraham E, Soffer A, Aurbach D. The rate-determining step of electroadsorption processes into nanoporous carbon electrodes related to water desalination. J Phys ChemC 2009;113(51):21319–27.

[17] Suss ME, Porada S, Sun X, Biesheuvel PM, Yoon J, Presser V. Water desalination via capacitive deionization: what is it and what can we expect from it? Energy Environ Sci 2015;8(8):2296–319.

[18] Zhao R, Biesheuvel PM, Miedema H, Bruning H, van der Wal A. Charge efficiency: a functional tool to probe the double-layer structure inside of porous electrodes and application in the modeling of capacitive deionization. J Phys Chem Lett 2010;1(1):205–10.

[19] Lee JY, Seo SJ, Yun SH, Moon SH. Preparation of ion exchanger layered electrodes for advanced membrane capacitive deionization (MCDI). Water Res 2011;45(17):5375–80.

[20] Hassanvand A, Wei KJ, Talebi S, Chen GQ, Kentish SE. The role of ion exchange membranes in membrane capacitive deionisation. Membranes 2017;7(3):54.

[21] Ohshima H, Kondo T. Membrane-potential and Donnan potential. Biophys Chem 1988;29(3):277–81.

[22] Ahualli S, Mar Fernandez M, Iglesias G, Jimenez ML, Liu F, Wagterveld M, Delgado AV. Effect of solution composition on the energy production by capacitive mixing in membrane-electrode assembly. J Phys Chem C 2014;118(29):15590–9.

[23] Makino K, Ohshima H, Kondo T. Surface-potential of an ion-penetrable charged membrane. J Theoret Biol 1987;125(3):367–8.

[24] Kwak NS, Koo JS, Hwang TS, Choi EM. Synthesis and electrical properties of NaSS-MAA-MMA cation exchange membranes for membrane capacitive deionization [MCDI]. Desalination 2012;285:138–46.

[25] Kang KW, Hwang CW, Hwang TS. Synthesis and properties of sodium vinylbenzene sulfonate-grafted poly(vinylidene fluoride) cation exchange membranes for membrane capacitive deionization process. Macromol Res 2015;23(12):1126–33.

[26] Jeong KS, Hwang WC, Hwang TS. Synthesis of an aminated poly(vinylidene fluride-g-4-vinyl benzyl chloride) anion exchange membrane for membrane capacitive deionization(MCDI). J Membr Sci 2015;495:316–21.

[27] Qiu Q, Cha JH, Choi YW, Choi JH, Shin J, Lee YS. Preparation of stable polyethylene membranes filled with crosslinked sulfonated polystyrene for membrane capacitive deionization by gamma-irradiation. Macromol Res 2017;25(1):92–5.

[28] Li HB, Zou L. Ion-exchange membrane capacitive deionization: a new strategy for brackish water desalination. Desalination 2011;275(1-3):62–6.

[29] Zhao YJ, Wang Y, Wang RG, Wu YF, Xu SC, Wang JX. Performance comparison and energy consumption analysis of capacitive deionization and membrane capacitive deionization processes. Desalination 2013;324:127–33.

[30] Ding M, Shi WH, Guo L, Leong ZY, Baji A, Yang HY. Bimetallic metal-organic framework derived porous carbon nanostructures for high performance membrane capacitive desalination. J Mater Chem A 2017;5(13):6113–21.

[31] Huyskens C, Helsen J, de Haan AB. Capacitive deionization for water treatment: screening of key performance parameters and comparison of performance for different ions. Desalination 2013;328:8–16.

[32] Choi JH. Comparison of constant voltage (CV) and constant current (CC) operation in the membrane capacitive deionisation process, Desalin. Desalin Water Treat 2015;56(4): 921–8.

[33] Zhao R, Porada S, Biesheuvel PM, Van der Wal A. Energy consumption in membrane capacitive deionization for different water recoveries and flow rates, and comparison with reverse osmosis. Desalination 2013;330:35–41.

[34] Tang WW, He D, Zhang CY, Waite TD. Optimization of sulfate removal from brackish water by membrane capacitive deionization (MCDI). Water Res 2017;121:302–10.

[35] Chen ZL, Zhang HT, Wu CX, Wang YS, Li W. A study of electrosorption selectivity of anions by activated carbon electrodes in capacitive deionization. Desalination 2015;369:46–50.

[36] Pan JF, Zheng Y, Ding JC, Gao CJ, Van der Bruggen B, Shen JN. Fluoride removal from water by membrane capacitive deionization with a monovalent anion selective membrane. Ind Eng Chem Res 2018;57(20):7048–53.

[37] Tang W, Kovalsky P, Cao B, He D, Waite TD. Fluoride removal from brackish groundwaters by constant current capacitive deionization (CDI). Environ Sci Technol 2016;50(19):10570–9.

[38] Yeo JH, Choi JH. Enhancement of nitrate removal from a solution of mixed nitrate, chloride and sulfate ions using a nitrate-selective carbon electrode. Desalination 2013;320:10–6.

[39] Kim YJ, Kim JH, Choi JH. Selective removal of nitrate ions by controlling the applied current in membrane capacitive deionization (MCDI). J Membr Sci 2013;429:52–7.

[40] Huang X, He D, Tang WW, Kovalsky P, Waite TD. Investigation of pH-dependent phosphate removal from wastewaters by membrane capacitive deionization (MCDI). Environ Sci Water Res Techno 2017;3(5):875–82.

[41] Mikhaylin S, Bazinet L. Fouling on ion-exchange membranes: classification, characterization and strategies of prevention and control. Adv Colloid Interf Sci 2016;229:34–56.

[42] Mossad M, Zou L. Study of fouling and scaling in capacitive deionisation by using dissolved organic and inorganic salts. J Hazard Mater 2013;244:387–93.

[43] Lee JB, Park KK, Yoon SW, Park PY, Park KI, Lee CW. Desalination performance of a carbon-based composite electrode. Desalination 2009;237(1-3):155–61.

[44] Liu Y, Pan LK, Xu XT, Lu T, Sun Z, Chua DHC. Enhanced desalination efficiency in modified membrane capacitive deionization by introducing ion-exchange polymers in carbon nanotubes electrodes. Electrochim Acta 2014;130:619–24.

[45] Kim JS, Jeon YS, Rhim JW. Application of poly(vinyl alcohol) and polysulfone based ionic exchange polymers to membrane capacitive deionization for the removal of mono- and divalent salts. Sep Purif Technol 2016;157:45–52.

[46] Kim YJ, Choi JH. Improvement of desalination efficiency in capacitive deionization using a carbon electrode coated with an ion-exchange polymer. Water Res 2010;44(3):990–6.

[47] Gao X, Omosebi A, Landon J, Liu KL. Surface charge enhanced carbon electrodes for stable and efficient capacitive deionization using inverted adsorption-desorption behavior. Energy Environ Sci 2015;8(3):897–909.

[48] Decher G, Hong JD, Schmitt J. Buildup of ultrathin multilayer films by a self-assembly process. 3. Consecutively alternating adsorption of anionic and cationic polyelectrolytes on charged surfaces. Thin Solid Films 1992;210(1-2):831–5.

[49] Decher G. Fuzzy nanoassemblies: toward layered polymeric multicomposites. Science 1997;277(5330):1232–7.

[50] Ahualli S, Iglesias GR, Fernandez MM, Jimenez ML, Delgado AV. Use of soft electrodes in capacitive deionization of solutions. Environ Sci Technol 2017;51(9):5326–33.

[51] Ahualli S, Jimenez ML, Fernandez MM, Iglesias G, Brogioli D, Delgado AV. Polyelectrolyte-coated carbons used in the generation of blue energy from salinity differences. Phys Chem Chem Phys 2014;16(46):25241–6.

[52] Gao X, Omosebi A, Holubowitch N, Liu A, Ruh K, Landon J, Liu K. Polymer-coated composite anodes for efficient and stable capacitive deionization. Desalination 2016;399:16–20.

[53] Jeon SI, Park HR, Yeo JG, Yang S, Cho CH, Han MH, Kim DK. Desalination via a new membrane capacitive deionization process utilizing flow-electrodes. Energy Environ Sci 2013;6(5):1471–5.

[54] Porada S, Weingarth D, Hamelers HVM, Bryjak M, Presser V, Biesheuvel PM. Carbon flow electrodes for continuous operation of capacitive deionization and capacitive mixing energy generation. J Mater Chem A 2014;2(24):9313–21.

[55] Hatzell KB, Boota M, Gogotsi Y. Materials for suspension (semi-solid) electrodes for energy and water technologies. Chem Soc Rev 2015;44(23):8664–87.

[56] Park HR, Choi J, Yang S, Kwak SJ, Jeon SI, Han MH, Kim DK. Surface-modified spherical activated carbon for high carbon loading and its desalting performance in flow-electrode capacitive deionization. RSC Adv 2016;6(74):69720–7.

[57] Doornbusch GJ, Dykstra JE, Biesheuvel PM, Suss ME. Fluidized bed electrodes with high carbon loading for water desalination by capacitive deionization. J Mater Chem A 2016;4(10):3642–7.

Section F

Conclusions

Chapter 9

Future and Perspectives of the Capacitive Techniques

Silvia Ahualli and Ángel V. Delgado
Department of Applied Physics, University of Granada, Granada, Spain

9.1 Introduction

In this book we have tried to give the reader an impression of the possibilities and achievements, but also the limitations, of capacitive techniques used for two objectives, reciprocal in some ways: energy harvesting and salt solution (partial) desalination. The two of them have been proven feasible, have received a good amount of theoretical justification, and have been implemented in a number of increasingly sophisticated ways [1–8]. They share the specific properties of electrical double layers (EDLs): they are capable of adsorbing a significant number of ions (that is, capable of being charged) if some electrical potential is applied to a pair of conducting interfaces (electrodes), and if the surface area is significant. This can be achieved by using activated carbon, with specific surface areas in the vicinity or above $1000\,m^2/g$, although the variety of tested materials is certainly abundant [9–18].

Energy can be obtained based on this principle by charging the EDLs in salt water by connecting the electrodes to an external power source, and discharging them in fresh water, after exchanging both solutions in open circuit conditions (constant charge). The key point is that the capacitance of the EDL is lower in fresh water and hence the potential is higher for given charge. Charge is received at high potential, and injected at low potential: as a result a net gain of energy is achieved as long as the solution exchange is sufficiently efficient. The method is denominated capacitive energy extraction based on double layer expansion (CDLE) [19]. The principle is thus easy, and it opens the possibility of using the well-established knowledge of interfaces for a socially significant target: obtaining clean energy, with absolutely no atmospheric or water contamination.

It is straightforward to understand how the same principles can be applied to deionization of aqueous electrolyte solutions: the ions (one charge sign on each electrode) accumulated in the EDLs are withdrawn from the solution in

Interface Science and Technology, Vol. 24. https://doi.org/10.1016/B978-0-12-811370-7.00009-7

between the electrodes, and thus this solution becomes (partially) deionized [10, 20, 21]. The cycle is concluded by discharging the electrodes from adsorbed ions (for instance, short-circuiting the electrodes or reversing the voltage) and discarding the resulting brine solution. Again, we understand the denomination given to the method: capacitive deionization (CDI).

9.2 Limitations of First-Principles Approaches

Simple as the methods appear, their limitations were soon diagnosed, even for such relatively young techniques. Regarding energy production, the capacitive cell shares the problem of self-discharge characteristic of the supercapacitor family [22, 23]: due to faradaic reactions, imperfect separation between electrodes, diffusion of adsorbed ions into the solutions, etc., the electrochemical capacitor loses its charge in a matter of hours typically. Brogioli et al. gave an additional argument, based on the tendency of the carbon surfaces to evolve towards their (low) surface charge [3]. This brought into play the coating of the electrodes in an attempt to stabilize the charge on them with no power source requirements [24]. In the case of CDI, the limitations are also apparent: the adsorption of counterions is accompanied by the desorption of coions, released in the solution.

9.3 Materials for Electrodes and Their Surface Treatment

Perhaps the main milestone in capacitive energy techniques was the use of ion exchange membranes (IEMs), cationic or anionic, on the corresponding electrode instead of (or, even better, in addition to) the use of an external power source. The membranes can charge the electrodes spontaneously by generation of membrane potential, while partially protecting them against self-discharge. The ratio between energy produced and spent is much more favorable. As a related alternative, the electrodes are coated with active compounds generating charge, very particularly polyelectrolytes, capable as well of producing a potential difference between the electrodes of Donnan origin, dependent on the ionic strength in contact with them [25–27].

The same applies to CDI: the use of ion exchange membranes on top of the active electrodes blocks the transport of the respective coions, and therefore for each electronic charge at the electrode one ion (counterion) is removed from solution, increasing in this way the charge efficiency of the process. The rate and amount of salt adsorption are certainly improved in MCDI.

It is also worth mentioning that the active materials themselves play an important role in the improvement of the performance of capacitive energy production. It has been shown [9, 18] that activated carbons can produce higher power if the pore size distribution is well designed, or when the hydrophilicity is favored, as both issues facilitate the exchange of ions in aqueous solution. Improvements have also been reported by changing the geometry, very significantly from flat to rod-shaped electrodes [28, 29]. The material selection has been more widely investigated in the case of CDI, where graphene, carbon nanotubes, or their combinations thereof have been investigated [13, 17, 30, 31].

9.4 Energy Balance: Where Are We Now?

We shall now consider the fundamental aspect of energy cost vs. benefits obtained. Let us first focus on the energy harvesting process: the question is: *how much power can we produce per unit mass of active material?* and *how much power do we have to spend for obtaining it?* We will consider the best conditions, which, according to the results of different groups, include [32–35]:

- Membrane- or polyelectrolyte-coated activated carbons
- Stacked cells, so that the scale is beyond the postal-stamp size
- Hierarchical carbons with well-balanced macro- and micropores

In such conditions, power densities on the order of 500 mW/m^2 at the most can be obtained, but typically the values are rather between 50 and 100 mW/m^2. Considering that the usual carbon load of an electrode is 300 g/m^2, the power density would be around 0.3 mW/g. The maximum value reported is about 3 mW/g for rod- or wire-shaped electrodes [36]. These figures do not take into account the amount of power lost because of self-discharge, and even if this could be minimized, one should still take into account the power required in pumping the solutions. A careful study was carried out by Veerman et al. [37–40] although particularized for reverse electrodialysis, so the information provided is only partially applicable to CapMix methods. Nevertheless, the hydraulic power evaluation can be of interest here. In [38], the authors found that for a stack of 50 cells, 100 cm^2 each, the hydraulic power loss was around 200 mW/m^2, or 25% of the total power produced, if the flow rate was 400 mL/min. If this fraction is also applicable to smaller-scale stacks, the decrease in power available might still be admissible.

In the case of CDI, a very complete analysis was recently reported by Suss et al. [8]. A primer quantity to be considered is salt adsorption capacity (SAC): the best values achieved reach 25 mg/g. This is not the only quantity of interest, as it is also important to ascertain the salt adsorption rate. Both quantities are plotted against each other in the so-called Kim-Yoon plots [41]; these authors demonstrated that it is possible to ensure both optimum SAC and adsorption rate (ASAR), and proposed a new parameter, called HCT (half-cycle time) for which the product SAC $\times$ ASAR is maximized. The second quantity, measuring how much energy we have to spend for each ion removed, is the charge efficiency, proportional to the inverse of the energy consumed per ion. Minimum energy spent is associated with maximum efficiency, and it is achieved in membrane or soft-electrode CDI [42, 43], reaching values down to

20 k_BT/ion. A related metric is *current efficiency*, or the ratio between salt adsorption rate and current through the cell.

As in energy harvesting, the selection of materials is also important in CDI. The list of proposed electrode materials during the last decade is rather long, and includes mainly carbon compounds, from simple activated carbon to graphene [10, 13, 16, 17, 44–46], nanotubes [17, 47–49], carbon aerogels [50–52], or composite electrodes. According to Suss et al. [8], a major issue is the stability and durability of the electrodes after repeated usage cycles, strongly dependent on the material used.

9.5 Economic Feasibility

It is likely that, once the technical issues discussed previously have been solved, economical considerations will also come into play. Regarding energy production, CapMix techniques would become acceptable from the point of view of the price per kilowatt-hour once it is possible to reach the target of 0.15 €/kWh at 2018 rates. Because of power losses, pumping and water pretreatment requirements, etc., it is unlikely that the power available is higher than 100 mW/m^2, with energy per cycle in the vicinity of 10 J/m^2. If the flow of water is 100 mL/min, and the cycle duration is around 1 min, the volume of river water exchanged in a cycle will be approximately 100 mL or 10^{-4} m^3. This means that the energy accumulated will be 1 kWh after exchanging 30 m^3 of river water. It sounds difficult to manage such large volumes at the price of a few euro cents, unless the power can be raised by at least one order of magnitude. There is considerable work ahead in order to improve these figures.

In the case of deionization, the fact that there is commercial interest from companies is proof that the method, particularly MCDI, is economically feasible. Thus, Voltea B.V. in The Netherlands (http:/voltea.com/en/) produces several pieces of equipment for applications ranging from cooling towers to laundry or agriculture. The Chinese company EST Water & Technologies has built large-scale facilities for wastewater treatment by CDI. Energy requirements for the equipment are in the vicinity of 1 kWh/m^3, a figure to be compared to that of reverse osmosis or flash distillation (4–10 kWh/m^3, see [53, 54]), for instance, although the latter technologies are suitable for seawater desalination, in contrast to capacitive ones.

9.6 Conclusion: Perspectives for the Future

It is difficult to predict how the capacitive techniques described in this book will evolve with time in the sense of improving their efficiencies to the limit of being comparable to technologies tested and installed almost universally at large scale. In principle, it might appear that it will be very difficult to match the achievements of more classical methodologies, but the fact is that improvements are still ahead in CapMix materials and methods, and that the latter can find niches of application (small scale, remote areas, developing communities, ...) where they are comparable to other methods in efficiency, and superior in ease of access and use of resources. When membranes or soft electrodes, as well as other architectures and materials, are properly combined with our advances in the fundamental knowledge, the possibilities of capacitive technologies will surely experience rapid growth.

Acknowledgments

Financial support for this work from Junta de Andalucía, Spain (Project PE2012-FQM-0694) is gratefully acknowledged. S.A. acknowledges the support from the University of Granada ("Vicerrectorado de Investigación, Plan Incorporación Jóvenes Doctores").

References

[1] Brogioli D. Extracting renewable energy from a salinity difference using a capacitor. Phys Rev Lett 2009;103(5):058501.

[2] Brogioli D, Zhao R, Biesheuvel PM. A prototype cell for extracting energy from a water salinity difference by means of double layer expansion in nanoporous carbon electrodes. Energy Environ Sci 2011;4(3):772–7.

[3] Brogioli D, Ziano R, Rica RA, Salerno D, Kozynchenko O, Hamelers HVM, et al. Exploiting the spontaneous potential of the electrodes used in the capacitive mixing technique for the extraction of energy from salinity difference. Energy Environ Sci 2012;5(12):9870–80.

[4] Brogioli D, Ziano R, Rica RA, Salerno D, Mantegazza F. Capacitive mixing for the extraction of energy from salinity differences: survey of experimental results and electrochemical models. J Colloid Interface Sci 2013;407:457–66.

[5] Rica RA, Ziano R, Salerno D, Mantegazza F, van Roij R, Brogioli D. Capacitive mixing for harvesting the free energy of solutions at different concentrations. Entropy 2013;15(4): 1388–407.

[6] Jimenez ML, Fernandez MM, Ahualli S, Iglesias G, Delgado AV. Predictions of the maximum energy extracted from salinity exchange inside porous electrodes. J Colloid Interface Sci 2013;402:340–9.

[7] Biesheuvel PM, Porada S, van der Wal A, Presser V. Carbon nanomaterials for water desalination by capacitive deionization. In: Gogotsi Y, Presser V, editors. Carbon Nanomaterials. 2nd ed. Boca Raton: CRC Press; 2014. ISBN: 9781439897812. pp. 419–60.

[8] Suss ME, Porada S, Sun X, Biesheuvel PM, Yoon J, Presser V. Water desalination via capacitive deionization: what is it and what can we expect from it? Energy Environ Sci 2015;8(8):2296–319.

[9] Iglesias GR, Fernandez MM, Ahualli S, Jimenez ML, Kozynchenko OP, Delgado AV. Materials selection for optimum energy production by double layer expansion methods. J Power Sources 2014;261:371–7.

[10] Li HB, Zou LD, Pan LK, Sun Z. Novel graphene-like electrodes for capacitive deionization. Environ Sci Technol 2010;44(22):8692–7.

[11] Xu YX, Sheng KX, Li C, Shi GQ. Self-assembled graphene hydrogel via a one-step hydrothermal process. ACS Nano 2010;4(7):4324–30.

[12] Lim HN, Huang NM, Lim SS, Harrison I, Chia CH. Fabrication and characterization of graphene hydrogel via hydrothermal approach as a scaffold for preliminary study of cell growth. Int J Nanomedicine 2011;6:1817–23.

[13] Wimalasiri Y, Zou LD. Carbon nanotube/graphene composite for enhanced capacitive deionization performance. Carbon 2013;59:464–71.

[14] Xu YX, Lin ZY, Zhong X, Huang XQ, Weiss NO, Huang Y, et al. Holey graphene frameworks for highly efficient capacitive energy storage. Nat Commun 2014;5:4554.

[15] Aphale A, Maisuria K, Mahapatra MK, Santiago A, Singh P, Patra P. Hybrid electrodes by in-situ integration of graphene and carbon-nanotubes in polypyrrole for supercapacitors. Sci Rep 2015;5:14445.

[16] Zhang YM, Zou LD, Wimalasiri Y, Lee JY, Chun Y. Reduced graphene oxide/polyaniline conductive anion exchange membranes in capacitive deionisation process. Electrochim Acta 2015;182:383–90.

[17] Jia BP, Zhang W. Preparation and application of electrodes in capacitive deionization (CDI): a state-of-art review. Nanoscale Res Lett 2016;11:64.

[18] Delgado AV, Ahualli S, Fernandez MM, Gonzalez MA, Iglesias GR, Vivo-Vilches JF, et al. Geometrical properties of materials for energy production by salinity exchange. Environ Chem 2017;14(5):279–87.

[19] Bijmans MFM, Burheim OS, Bryjak M, Delgado A, Hack P, Mantegazza, F, Tenisson S, Hamelers HVM, CAPMIX—deploying capacitors for salt gradient power extraction, techno-port 2012, In Sharing possibilities and 2nd renewable energy research conference (Rerc2012), vol. 20; 2012, pp. 108–115.

[20] Oren Y. Capacitive delonization (CDI) for desalination and water treatment—past, present and future (a review). Desalination 2008;228(1-3):10–29.

[21] Porada S, Weinstein L, Dash R, van der Wal A, Bryjak M, Gogotsi Y, et al. Water desalination using capacitive deionization with microporous carbon electrodes. ACS Appl Mater Interfaces 2012;4(3):1194–9.

[22] Conway BE. Electrochemical supercapacitors. Scientific fundamentals and technological applications. New York: Plenum Press; 1999.

[23] Andreas HA. Self-discharge in electrochemical capacitors: a perspective article. J Electrochem Soc 2015;162(5):A5047–53.

[24] Marino M, Misuri L, Jimenez ML, Ahualli S, Kozynchenko O, Tennison S, et al. Modification of the surface of activated carbon electrodes for capacitive mixing energy extraction from salinity differences. J Colloid Interface Sci 2014;436:146–53.

[25] Sales BB, Saakes M, Post JW, Buisman CJN, Biesheuvel PM, Hamelers HVM. Direct power production from a water salinity difference in a membrane-modified supercapacitor flow cell. Environ Sci Technol 2010;44(14):5661–5.

[26] Liu F, Schaetzle O, Sales BB, Saakes M, Buisman CJN, Hamelers HVM. Effect of additional charging and current density on the performance of capacitive energy extraction based on Donnan Potential. Energy Environ Sci 2012;5(9):8642–50.

[27] Ahualli S, Jimenez ML, Fernandez MM, Iglesias G, Brogioli D, Delgado AV. Polyelectrolyte-coated carbons used in the generation of blue energy from salinity differences. Phys Chem Chem Phys 2014;16(46):25241–6.

[28] Burheim OS, Liu F, Sales BB, Schaetzle O, Buisman CJN, Hamelers HVM. Faster time response by the use of wire electrodes in capacitive salinity gradient energy systems. J Phys Chem C 2012;116(36):19203–10.

[29] Sales BB, Burheim OS, Liu F, Schaetzle O, Buisman CJN, Hamelers HVM. Impact of wire geometry in energy extraction from salinity differences using capacitive technology. Environ Sci Technol 2012;46(21):12203–8.

[30] El-Deen AG, Boom RM, Kim HY, Duan HW, Chan-Park MB, Choi JH. Flexible 3D nanoporous graphene for desalination and biodecontamination of brackish water via asymmetric capacitive deionization. ACS Appl Mater Interfaces 2016;8(38):25313–25.

[31] Ahmed MA, Tewari S. Capacitive deionization: processes, materials and state of the technology. J Electroanal Chem 2018;813:178–92.

[32] Galindo-Gonzalez C, Iglesias GR, Gonzalez-Caballero F, Duran JDG. Stability of concentrated aqueous clay-magnetite suspensions. Colloids Surf A-Physicochem Eng Aspects 2007;306(1-3):150–7.

[33] Ahualli S, Fernandez MM, Iglesias G, Delgado AV, Jimenez ML. Temperature effects on energy production by salinity exchange. Environ Sci Technol 2014;48(20):12378–85.

[34] Iglesias GR, Ahualli S, Fernandez MM, Jimenez ML, Delgado AV. Stacking of capacitive cells for electrical energy production by salinity exchange. J Power Sources 2016;318: 283–90.

[35] Yip NY, Brogioli D, Hamelers HVM, Nijmeijer K. Salinity gradients for sustainable energy: primer, progress, and prospects. Environ Sci Technol 2016;50(22):12072–94.

[36] Liu F, Advancing CapMix for electricity generation. Ph.D. thesis, University of Wageningen, Wageningen, The Netherlands; 2015.

[37] J. Veerman, Reverse electrodialysis. Design and optimization by modeling and experimentation. Ph.D. thesis, University of Groningen, The Netherlands, Rijksuniversiteit Groningen, Groningen; 2010.

[38] Veerman J, Saakes M, Metz SJ, Harmsen GJ. Reverse electrodialysis: performance of a stack with 50 cells on the mixing of sea and river water. J Membr Sci 2009;327(1-2): 136–44.

[39] Veerman J, Saakes M, Metz SJ, Harmsen GJ. Reverse electrodialysis: a validated process model for design and optimization. Chem Eng J 2011;166(1):256–68.

[40] Tufa RA, Pawlowski S, Veerman J, Bouzek K, Fontananova E, di Profio G, et al. Progress and prospects in reverse electrodialysis for salinity gradient energy conversion and storage. Appl Energy 2018;225:290–331.

[41] Kim T, Yoon J. CDI ragone plot as a functional tool to evaluate desalination performance in capacitive deionization. RSC Adv 2015;5(2):1456–61.

[42] Zhao YJ, Wang Y, Wang RG, Wu YF, Xu SC, Wang JX. Performance comparison and energy consumption analysis of capacitive deionization and membrane capacitive deionization processes. Desalination 2013;324:127–33.

[43] Ahualli S, Iglesias GR, Fernandez MM, Jimenez ML, Delgado AV. Use of soft electrodes in capacitive deionization of solutions. Environ Sci Technol 2017;51(9):5326–33.

[44] Li HB, Zou L, Pan LK, Sun Z. Using graphene nano-flakes as electrodes to remove ferric ions by capacitive deionization. Sep Purif Technol 2010;75(1):8–14.

[45] Jia BP, Zou LD. Graphene nanosheets reduced by a multi-step process as high-performance electrode material for capacitive deionisation. Carbon 2012;50(6):2315–21.

[46] Yan C, Kanaththage YW, Short R, Gibson CT, Zou LD. Graphene/polyaniline nanocomposite as electrode material for membrane capacitive deionization. Desalination 2014;344:274–9.

[47] Chen JH, Li WZ, Wang DZ, Yang SX, Wen JG, Ren ZF. Electrochemical characterization of carbon nanotubes as electrode in electrochemical double-layer capacitors. Carbon 2002;40(8):1193–7.

[48] Yan CJ, Zou L, Short R. Single-walled carbon nanotubes and polyaniline composites for capacitive deionization. Desalination 2012;290:125–9.

[49] Liu Y, Pan LK, Xu XT, Lu T, Sun Z, Chua DHC. Enhanced desalination efficiency in modified membrane capacitive deionization by introducing ion-exchange polymers in carbon nanotubes electrodes. Electrochim Acta 2014;130:619–24.

[50] Kumar R, Sen Gupta S, Katiyar S, Raman VK, Varigala SK, Pradeep T, et al. Carbon aerogels through organo-inorganic co-assembly and their application in water desalination by capacitive deionization. Carbon 2016;99:375–83.

[51] Rasines G, Lavela P, Macias C, Zafra MC, Tirado JL, Ania CO. Mesoporous carbon black-aerogel composites with optimized properties for the electro-assisted removal of sodium chloride from brackish water. J Electroanal Chem 2015;741:42–50.

[52] Rasines G, Lavela P, Macias C, Zafra MC, Tirado JL, Parra JB, et al. N-doped monolithic carbon aerogel electrodes with optimized features for the electrosorption of ions. Carbon 2015;83:262–74.

[53] Malaeb L, Ayoub GM. Reverse osmosis technology for water treatment: state of the art review. Desalination 2011;267(1):1–8.

[54] Ihm S, Al-Najdi OY, Hamed OA, Jun G, Chung H. Energy cost comparison between MSF, MED and SWRO: Case studies for dual purpose plants. Desalination 2016;397:116–25.

Index

Note: Page numbers followed by "*f*" indicate figures, and "*t*" indicate tables.

A

Activated carbon (AC), 65
Advancing angle, 130–131, 132*t*, 132*f*
Amphiphilic molecules, 71
Anion-exchange membrane (AEM), 176–177, 177*f*

B

Batch-pass mode, 171*f*, 172
BET. *See* Brunauer Emmett Teller (BET)
Bikerman equation, 40, 43–44
Blue energy, 122, 133–134
Boltzmann distribution, 4–5, 29, 150–151
Born force, 56
Boublik-Mansoori-Carnahan-Starling-Leland (BMCSL), 40, 44, 51–52, 54
Brunauer Emmett Teller (BET), 39–41, 50, 57–58, 66–67

C

Capacitance, EDL
  activated carbon particles, 39–40
  Bikerman equation *vs.* C-S model, 44–46
  CDI (*see* Capacitive deionization (CDI))
  double-layer capacitors, 39
  electrode/solution interface, 40
  excess chemical potential, 43–44
  ionic adsorption mechanisms, 39
  ion specificity (*see* Ion specificity)
  mD approach, 42–43
  mean field theory, 40
  monovalent cations, 40
  multiionic solutions
    different valency, 51–52
    equal valency, 53–55
    size effects, steric repulsion, 56
  nonideal ionic behavior, 40
  overlap, 48–50
  steric effect, CDLE, 56–59
  steric repulsion contribution, 40
  surface charge and potential, 41–42, 54*f*
Capacitive deionization (CDI), 40, 195–196
  basic principles, 170–175
  electrochemical demineralization, 169–170
  electrodes, carbon materials
    carbon aerogels, 66–68
    CNTs, 68–70
    graphene, 73
    mesoporous carbon, 70–73
    nitrogen-doped carbons (*see* Nitrogen-doped carbons)
  flow electrodes, 185–187
  membrane capacitive deionization
    applications of, 181–182
    basic principles, 176–177
    practical issues, 180–181
  performance evaluation, 175–176
  reverse osmosis, 169
  techniques, 40
  treated electrodes, 182–185
Capacitive Donnan potential (CDP)
  efficiency, 161–164
  experimental implementation, 152–156
  ion exchange membrane, 144–145
  membrane potential, 143–144
  monovalent electrolyte, 144–145
  power, increase of
    additional charging, 157–160
    wire electrodes, stacking of, 161
  principle, 91–92
  self-discharge, 143
  soft electrodes, 149–151
  wire electrodes, 156–157
Capacitive double layer expansion (CDLE)
  Capmix cells
    activated carbon films, 120–121
    blue energy, 122
    Coulombic interactions, 121
    diffusion, 121
    specific surface area, 121, 121*t*
    stackings of, 135–138
    Stern layer, 121
    wettability, carbon samples, 121
  CapMix cycle, 94–96, 95*f*

Capacitive double layer expansion (CDLE) (*Continued*)
carbon materials requirements, 130–133
charge leakage, 119
energy harvesting cycle, 26
kinetic aspects of, 110–112
methodology, 122–124
optimum working conditions, 124–127
salinity difference effects, 133–135
salinity gradient power, 87–92
CapMix technique advantages, 91–92
charge storage, porous electrodes, 90–91
desalination battery, 91–92
environmental footprint, 87
free energy dissipation, 87–88
membrane-CapMix, 91–92
mixing entropy battery, 90–91
pressure-retarded osmosis technique, 88, 88*f*
pretreatments, 89
reverse electrodialysis technique (RED), 88–89, 89*f*
salt adsorption theory, 96–102
supercapacitors, 120
surface charge density, 119
temperature variations, 119–120
thermodynamics, capmix process, 102–104
voltage rise
activated carbon, surface modifications, 107–110
base voltage graph, 104–105
Capacitive energy extraction
CDLE (*see* Capacitive double layer expansion (CDLE))
Donnan potential approach, 144
Capacitive techniques
economic feasibility, 198
electrodes and surface treatment, 196–197
energy balance, 197–198
first-principles approaches, 196
perspectives, 198–199
Carbon aerogels (CAs), 65–68
Carbon materials, CDI electrodes
carbon aerogels, 66–68
CNTs (*see* Carbon nanotubes (CNTs))
graphene, 73
mesoporous carbon, 70–73
nitrogen-doped carbons (*see* Nitrogen-doped carbons)
Carbon nanotubes (CNTs), 65, 68–70
Carnahan-Starling (C-S) equation, 40
CAs. *See* Carbon aerogels (CAs)
Catalytic chemical vapor deposition (CCVD), 69
Cation-exchange membrane (CEM), 176–180, 177*f*
CDI. *See* Capacitive deionization (CDI)
Cell potential, 26–28, 26*f*, 129, 129*f*
Charge balance equation, 110–111
Charge density, 105, 108
Charge efficiency, 101, 176
Concentration capacitor, 152–153, 153*f*
Constant current (CC), 124, 172, 173*f*
Constant resistance, 33
Constant voltage (CV), 123, 172, 173*f*

D

Debye-Hückel approximation, 6
Debye length, 5–6, 12, 48–49, 93
Desalination charge efficiency, 103, 105
Dielectrophoretic force, 56
Diffuse potential, 25
Diffusion coefficient, 32–33
Dissociation constants, 8
Donnan equilibrium, 32–34
Donnan potential, 12–13, 34, 88–89
Double layer permittivity exchange (DLPE), 127–129

E

Electrical double layers (EDLs), 65, 119, 143, 195–196
capacitance (*see* Capacitance, EDL)
CDLE, 92–94
porous materials models
capacitance of, 3–4
charge-potential relationships, 4
cylindrical pores, 8–12
energy storage devices, 3–4
Gouy-Chapman-Stern-Grahame model, 6–8
Gouy-Chapman theory, 4–6
Helmholtz model, 3, 4*f*
modified-Donnan model, 12–13
simulations, 13–16
supercapacitor, 3–4
Electrode specific resistance, 34
Equilibrium constants, 8

F

Faradaic reactions, 124, 196
Fick's law, 21–22, 156–157
Flow-electrode capacitive deionization (FCDI), 185, 186*f*
Flow-through electrode, 174
Fluidized bed electrodes (FbedCDI), 187

G

Gibbs-Duhem equation, 98
Gibbs free energy, 103
Gouy, Chapman, and Stern (GCS) theory, 13–14, 13–14*f*, 105–106, 110
Graphene, 73
Graphene hydrogels (GH), 152–153, 154*f*
Graphene oxide sponge (GOS), 78–79
Graphene sponge (GS), 78
Graphite oxide (GO), 77

H

Half-cycle time (HCT), 197–198
Helmholtz model, 3–5, 4*f*
Hummer's method, 73
Hypernetted chain (HNC), 14

I

IEMs. *See* Ion exchange membranes (IEMs)
Inner Helmholtz plane (IHP), 7–8
Integral equation approach, 14
Inverted capacitive deionization (I-CDI), 182–183
Ion chromatography, 171–174
Ion exchange membranes (IEMs), 176–183, 185, 196
Ionic diffusion coefficient, 22
Ionic flux, 32–33
Ions and salt adsorption kinetics, porous electrodes
  energy production, implications for, 26–27
  equations and boundary conditions, 21–24
  faradaic reactions, 19
  macropores, 19, 20*f*, 21, 30, 33
    ion adsorption, 25
    ionic fluxes, 30
    mass balance equation, 32
    modified Donnan model, 33
    salt concentration, 27, 34–35, 35*f*
    surface charge density, 23
    voltage drop, 23
  MCDI (*see* Membrane capacitive deionization (MCDI))
  membrane capacitive deionization (MCDI), 26–27, 31–36
  micropores, 19, 20*f*, 24, 30, 33
    diffuse and Stern potential drops in, 23
    ionic fluxes, 30
    mass balance equation, 32
    modified Donnan model, 24, 33
    salt adsorption/desorption, 22
    spacer concentration, 28
    Stern layer, 23–24
    volume charge density, 22–24
  1-D model, 28, 31
  time, concentration and potential profiles with, 24–25
Ion specificity
  ionic size, effect of, 48
  valency, effect of, 47–48

L

Langmuir-type adsorption isotherm, 8
Lennard-Jones potential, 15–16

M

Mass balance equation, 29, 32, 110–111
Mean spherical approximation (MSA), 14
Melamine (M), resorcinol (R) and formaldehyde (F) (MRF), 73–75, 76–77*f*
Membrane capacitive deionization (MCDI), 31–35
  applications of, 181–182
  basic principles, 176–177
  practical issues, 180–181
Mesoporous carbon (MC), 65, 70–73
Modified Donnan (mD), 24, 29, 33, 42–43
Monte Carlo (MC) simulations, 13–14, 13–14*f*
Multiionic (MI) solutions
  different valency, 51–52
  equal valency, 53–55
  size effects, steric repulsion, 56

N

Nernst-Planck equations, 21–22, 29, 32–33
Nitrogen-doped carbons, carbon aerogels, 73–76
Nitrogen-doped graphene (NG), 77–79
Nitrogen-doped graphene sponge (NGS), 78–79, 78*f*
Nitrogen-doped porous carbon spheres (NPCSs), 79–81

O

Open-circuit voltage (OCV), 156–157
Ordered mesoporous carbons (OMCs), 71–72
Ornstein-Zernike formulation, 14
Outer Helmholtz plane (OHP), 6–7

P

PCSs. *See* Porous carbon spheres (PCSs)
Point-like ion model (PI), 43

Poisson-Boltzmann equation, 6, 8–9
Polyelectrolyte coatings, 149, 150*f*, 155–156
Polyethyleneimine branched with ethylene diamine (PEI-EN), 109
Pore size distribution, 130–133, 131*f*
Porous carbon spheres (PCSs), 79–81
Potential of zero charge (PZC), 68
Pristine graphene (PG), 77

R

Reduced graphene (RG) sheets, 77
Reverse current (RC), 172
Reverse electrodialysis (RED), 88–91
Reverse osmosis (RO), 169
Reverse voltage (RV), 172, 173*f*, 177

S

Salinity gradient power (SGP), 87–92
  CapMix technique advantages, 91–92
  charge storage, porous electrodes, 90–91
  desalination battery, 91–92
  environmental footprint, 87
  free energy dissipation, 87–88
  membrane-CapMix, 91–92
  mixing entropy battery, 90–91
  pressure-retarded osmosis technique, 88, 88*f*
  pretreatments, 89
  reverse electrodialysis technique (RED), 88–89, 89*f*
Salt adsorption capacity (SAC), 175, 197–198
Salt diffusion coefficient, 156–157
Salt removal efficiency, 180*f*
Self-discharge phenomenon, 105–107, 109
SGP. *See* Salinity gradient power (SGP)
Single-pass mode, 171*f*, 172
Soft electrodes (SEs), 149–151, 155*f*
  CDI, soft electrode assembly, 30, 31*f*
  CDP, 149–151
Spacer channel, 34
Stagnant diffusion layer (SDL), 20–21, 26–28
Stern layer, 6–8, 7*f*, 10–13, 42–43
Stern volume capacitance, 26–28
Surface charge density, 5–8, 13*f*, 23, 41, 44, 50*f*, 92, 134
Surface potential, 9, 13–14

T

Thermal membrane potential, 129
Thermal power plants, 127
Thermodynamic efficiency, 161–162

V

Voltage-charge cycles, 157–158, 162
Volume charge density, 24, 42–43

W

Wire electrodes, 156–157

Z

Zero current (ZC), 172
Zero-voltage (ZV) mode, 172, 173*f*

FSC
www.fsc.org
MIX
Papier | Fördert
gute Waldnutzung
FSC® C083411

Kolibri 360 GmbH
Ferdinand-Jühlke-Straße 7
99095 Erfurt, Deutschland
produktsicherheit@kolibri360.de